Slippery Math in Public Affairs

Slippery Math in Public Affairs

Price Tag and Defense

William J. Adams
Pace University
New York, New York

with illustrations by
Ramunė B. Adams

MARCEL DEKKER, INC. NEW YORK • BASEL

ISBN: 0-8247-0790-7

This book is printed on acid-free paper.

Headquarters
Marcel Dekker, Inc.
270 Madison Avenue, New York, NY 10016
tel: 212-696-9000; fax: 212-685-4540

Eastern Hemisphere Distribution
Marcel Dekker AG
Hutgasse 4, Postfach 812, CH-4001 Basel, Switzerland
tel: 41-61-261-8482; fax: 41-61-261-8896

World Wide Web
http://www.dekker.com

The publisher offers discounts on this book when ordered in bulk quantities. For more information, write to Special Sales/Professional Marketing at the headquarters address above.

Current printing (last digit):
10 9 8 7 6 5 4 3 2 1

PRINTED IN THE UNITED STATES OF AMERICA

To
Onutė, Andrius, Barbara, Brian, Ramunė, Gaja, Rasa, and Roksi

Preface

Plato believed that knowledge of mathematics is a prerequisite for good citizenship. In our age it would be appropriate to consider knowledge of how slippery math enters the public arena and what we can do to avoid being "taken" to be a prerequisite for effective participation in public affairs.

In brief, this book addresses this post-Plato state of affairs. In doing so, it presupposes no specialized knowledge of mathematics on the part of the reader, just interest in how mathematics may deceive us, the cost, and what we can do about it.

Slippery math in public affairs? At first the thought might strike one as an oxymoron. After all, many of us would submit that in terms of precision, mathematics is as good as it gets. We acquire such a view of mathematics at an early age and it stays with us, although with further education and life experience we come to realize that math is not quite all that it's cracked up to be. In a sense math is like a respectable public figure whom we are drawn to because he projects strength, objectivity, and logical power, but whom we also sense has a shadowy side we can't quite pinpoint.

Our degree of vulnerability to being seduced by the Svengali-like presence of mathematics depends on the faith and trust we have in its power, balanced by our sense of caution and doubt. Alas, the hypnotic power of mathematics far too often scores a complete triumph. This was brought home to me by the reaction of a student to an argument I had presented concerning a dean who had been told that his work was counterproductive. The argument purported to show, by "mathematical proof," that the dean did no work at all, and thus that the charge had no basis. The point I was attempting to make was that "mathematical proof" could generate nonsense and that if it did so one would be well advised to take a close look at the mathematical proof and its starting point. My student Laura would have none of it; I had "proved" the result by mathematical reasoning and that was that; case closed. It's enough to make one want to laugh and cry at the same time. Although Laura might seem an extreme case, the fact is that we all have a bit of Laura in us.

The following questions warrant consideration:

1. What is slippery math and how does it arise?
2. What have been the consequences of slippery math in the public arena?
3. How is math employed to sell a point of view to the public?
4. With billions of dollars and endless talk given over to improving mathematics education, how could it happen that math can be employed to mislead most of the people most of the time?
5. What reforms in mathematics education are needed to help prepare future generations to participate effectively in a society in which mathematics is increasingly being employed and misemployed in the public arena?
6. What can we do to make ourselves less vulnerable to being manipulated by slippery math?
7. What can we do to help policy decision making reflect more reliable and well-chosen numbers and math models?

The setting for addressing these questions is established in Parts 1 through 4. We return to them in the Epilogue. Part 1 (Chapters 1–7) addresses the ways in which slippery numbers emerge in the public arena and what steps we can take to make ourselves less vulnerable to being math manipulated. Part 2 (Chapters 8–13) is concerned with the development of math portraits, called math models, for real-world situations, how they can be mathematically precise and yet present a false picture, the spin doctor's bag of tools to sell math portraits to the public, and what we can do to minimize the danger of being sold math snake oil. Part 3 (Chapters 14–15) discusses the cost of slippery math in terms of resources, life, and opportunities. Part 4 (Chapter 16) looks at how math education can be reformed to sharpen our perspective on what mathematics can and cannot do for us in its applications, and options available to us for putting together a self-study program.

I should like to express my appreciation to my daughter Ramunė for her illustrations, and to my daughter Rasa, Ursula Shand, Danielle Ortiz, and Eliana Cifuentes for their assistance in preparing the manuscript. I also should like to express my appreciation to the Pace University community for their support of this undertaking.

William J. Adams

Contents

Part One

SLIPPERY NUMBERS IN THE PUBLIC ARENA

INTRODUCTION

Words such as *accurate, precise, objective,* and *logical* derive their power by being symbols of an ideal information state from which ambiguity, ignorance, and prejudice have been banished. In our age mathematics is seen as the epitome of this information state to the degree that the respectability of a presentation in the conduct of public affairs is seen as enhanced, if not unequivocally defined, by numbers, statistics, and "mathematical proof." For numbers and statistics this is due to the acceptance by policy makers and the public at large of one or more of the following views.

1. A presentation supported by figures is more credible than one that is not.
2. Numbers are neutral. They are not affected by cultural, economic, and political differences that influence people and therein lies their strength. $1 + 1 = 2$, for example, is universal for all peoples, countries, economic and political systems.
3. Numbers give weight to a view through the sense of precision they communicate, thereby advancing it to a plane which commands recognition, respect, and acceptance.

Do these views express math insights, math myths, or a bit of both attractively packaged as mathematical trojan horses? This is one of the questions addressed in Part One.

Which is preferable, numbers which lend to a study a sense of substance and precision, but are wide of reality's mark, or no numbers at all? If you agree with Artemus Ward that "it ain't so much the things we don't know that get us in trouble, it's the things we know that ain't so," then no numbers at all would be the preferable alternative. Since this state of affairs is not a viable option in our quantitatively oriented age, our only real alternative is to develop a sensitivity which would enable us to distinguish numbers that are wide of reality's mark from those that are reliable and insightful.

Part One addresses this task in the first seven chapters. Different dimensions of the reliability of tantalizing numbers that are floated our way are taken up in Chapters 1 to 3. Political pressures and ideological commitments are strong external forces that have a powerful affect on the reality of numbers we see and decisions that are ultimately taken. This dimension of number pollution is addressed in Chapter 4. The call to get the numbers is heard when a study is undertaken, but which numbers would be most "appropriate" for the study envisioned? This is often not clear, and there is no quicker way to undermine a study than to focus on numbers which are peripheral to the heart of the issue at hand. Chapter 5 is concerned with the question of how well chosen the numbers that have been taken as a foundation stone for a study are. Numbers by themselves are neutral but, like tea leaves, individuals and interest groups read them in different ways. The question is, which interpretation is "closest" to reality's mark? Chapter 6 takes up the interpretive dimension of numbers.

With a perspective obtained from Chapters 1 to 6, Chapter 7 returns to the aforenoted views on numbers. It develops a series of test questions that may be applied to a presentation founded on numbers and statistics to help us evaluate the credibility of the numerical support system.

1

Boldfaced Lies

As the following cases illustrate, boldfaced numerical lies come in many sizes, shapes, and forms. They may arise from any source at any time. It's not a matter of the background activity involved, but the honesty and integrity of the people who are part of the activity. Our best protection against the dissemination of boldfaced numerical lies and the damage they may cause is to have an appropriate system of checks and balances.

Boldfaced numerical lies usually arise from a desire to further a "worthy" cause or simple self-interest.

THE HOMELESS HOMELESS FIGURE

In their 1982 book, *Homelessness in America: A Forced March to Nowhere,* advocates for the homeless Mary Hombs and Mitch Snyder estimated that between 2 and 3 million people in the United States were homeless, suggesting that on the order of 1 out of 100 Americans had no place to live. This estimate seemed plausible to many residents of large cities in terms of anecdotal evidence they had been exposed to, but a few years after citing this estimate Snyder admitted that he had made up this number to alleviate pressures to be specific [1].

INDEFENSIBLE DEFENSE FIGURES

In 1992 the General Accounting Office audited seven "Stars Wars" tests conducted between 1990 and 1992. It found that four of the test results described to Congress as successes were false, whereas the three tests that were described as complete or partial failures were correct [2].

Are recent antimissile test results more reliable? I wouldn't count on it. Dr. Nira Schwartz, a former engineer who worked on designing computer programs to enable interceptor missiles to distinguish between incoming warheads and decoys for the military contractor TRW, charged that key test results were deliberately faked. Dr. Schwartz claimed that while TRW certified that interceptors using its computer programs had better than a 95 percent success rate in distinguishing warheads from decoys, the actual success rate was 5 to 15 percent [3; see also Refs. 4–8].

BLOWING SMOKE ON SECONDARY SMOKE

An important support for the tobacco industry's argument against control of smoking in workplaces was provided by a study based on measurements of environmental tobacco smoke in a survey of 585 buildings in San Diego conducted by Healthy Buildings International (H.B.I.), a private company located in Virginia.

The problem is that over 25 percent of the measurements had significant alterations which undercounted the amount of cigarette smoke as measured by suspended particles in the air. One field technician, Gregory Wulchin, reported: "After I submitted my field notes to H.B.I., Gray Robertson, the H.B.I. president, changed the particular measurement I recorded for the sample. In his own handwriting, he struck out the number 150 in the field notes and inserted the number 75." In response, Mr. Robertson said that he had changed the reading because he guessed it was too high, considering that other measurements in the building were lower. He said he took the number, cut it in half, then in his final report estimated the figure slightly higher, at 85. Mr. Robertson also commented, in a telephone interview, that the methods he used in collecting and altering the data were common in the indoor-air-quality industry and were not fraudulent. The Tobacco Institute stated that it considered Healthy Buildings International "a fine firm."

Dr. Stanton, a cardiologist at the University of California at San Francisco, commented, "This data is just crucial. The argument of the tobacco companies has been that the exposure levels of environmental smoke in workplaces is so low as to be trivial. If this data is fraudulent, it throws into question all their assertions [9].

SHENANIGANS IN SCIENCE

Is there a link between electric power lines and cancer? This was a hot topic in the early 1990s, and two papers published in 1992 written by Dr.

Robert Liburdy, a cell biologist at the Lawrence Berkeley Laboratory in Berkeley, California, provided evidence that the answer is yes. "Now both the lab, and the Office of Research Integrity, have found that data on which he based his conclusions were fabricated," stated Glenn R. Woods, the laboratory's counsel, in an interview conducted in July 1999 [10].

THE DRUG TESTING BALLOON THAT POPPED

Dr. Robert Fiddes was on the top of the world. He had gone from successful medical practitioner to founder and director of the Southern California Research Institute, a company that tested the effectiveness and safety of new drugs for pharmaceutical companies and brought in millions of dollars. Slightly behind this stunning facade of success and prosperity, research fraud of almost unimaginable proportions was in progress.

It wasn't simply a matter of altering numerical readings, although this was done too, but of compromising the entire framework for obtaining meaningful readings. Blood pressure readings were indeed fabricated, but fictitious patients were invented, medical records were falsified, medical tests were compromised, patients who did not satisfy the criteria for inclusion in a study were enrolled anyway.

The balloon was punctured on February 16, 1997 when federal agents occupied the Southern California Research Institute's Office and confiscated box loads of incriminating documents, leaving in its wake compromised drug study results for almost every pharmaceutical house in the business [11].

YOUR DATA HAVE
TAKEN A TURN FOR
THE WORSE!

2

Something's Amiss

Boldfaced numerical lies, troublesome as they are, provide us with one dimension of number manipulation. Alas, the generation of numerical fiction takes many forms which do not lend themselves to simple classification. Some of these forms are considered here.

IT SOUNDS GOOD, BUT . . .

It's There on Paper, But . . .

In his state of the union message for 1999 President Bill Clinton announced a plan to devote $2.8 trillion of the $4.5 trillion projected budget surplus over the next 15 years to shoring up Social Security. By law, some $2.3 trillion of the $4.5 trillion is already committed to Social Security. This yields $5.1 trillion of an envisioned $4.5 trillion surplus that is being committed to Social Security, on the books at least. Is this double counting of $2.3 trillion a bookkeeping gimmick? Yes, says Congressional Budget Office Director Dan Crippen. No, says Federal Reserve Bank Director Alan Greenspan [1,2].

You Have It, But You Don't

In early 1981 Ronald Reagan's budget director David Stockman found himself drowning in a lake of red ink. President Reagan had promised that the country would have a balanced budget by 1983, and Stockman found himself forced to move the target date to 1984.

Stockman writes [3]:

> But that was merely a straw in the wind compared to what would come next. I soon became a veritable incubator of shortcuts, schemes, and devices to overcome the truth now upon us—that the budget gap couldn't be closed except by a dictator.
>
> The more I flopped and staggered around, however, the more they went along. I could have been wearing a sandwich board sign saying: Stop me, I'm dangerous! Even then they might not have done so . . .
>
> Bookkeeping invention thus began its wondrous works. We invented the "magic asterisk": If we couldn't find the saving in time—and we couldn't—we would issue an IOU. We would call it "Future savings to be identified."
>
> It was marvelously creative. A magic asterisk item would cost negative $30 billion . . . $40 billion . . . whatever it took to get a balanced budget in 1984 after we toted up all the individual budget cuts we'd actually approved.

The magic asterisk passed presidential and congressional muster.

A refinement on the magic asterisk device was employed by the Clinton team in proposing spending programs for the 1999 budget. The magic asterisk took the form of "to be funded by tobacco settlement money," funds that were iffy at the time and which the government never realized.

ACCOUNTING ALCHEMY: FROM LEAD TO GOLD

It was an unusual scene, with L. Dennis Kozlowski, CEO of what has been termed the most voracious acquisition machine in corporate America, Tyco International Ltd., playing defense. He was responding to allegations made by David W. Tice and Associates, a company that provides accounting related research to institutional investors, that the rosy growth picture painted by Tyco was based on accounting gimmicks [4,5]. Alas, this scenario, in its essentials, is far from unique.

> Many major corporations still play things straight, but a significant and growing number of otherwise high-grade managers . . . have come to the view that it's OK to manipulate earnings to satisfy what they believe are Wall Street's desires. These managers often say that their shareholders will be hurt if their currency for doing deals—that is, their stock—is not fully priced, and they also argue that in using accounting shenanigans to get the figures they want, they are only doing what everybody else does. Once such an everybody's-doing-it attitude takes hold, ethical misgivings vanish [6].

An attack from the Securities and Exchange Commission? No. Although the Securities and Exchange Commission, under chairman Arthur Levitt, has vigorously attacked the practice of earnings management through financial shenanigans, this concern was voiced by one of the most highly respected investors of our time, Warren E. Buffett. Buffett, like many investors, believes that for the investment market to function well ethical principles must be given highest priority and that honesty, indeed, is the best policy.

Howard M. Schilit defines financial shenanigans as "actions or omissions intended to hide or distort the real financial performance or financial condition of an entity" [7]. Schilit lists, discusses, and illustrates seven major shenanigans which enable management to manipulate net income.

S1: Recording revenue before it is earned

S2: Creating fictitious revenue

S3: Boosting profits with nonrecurring transactions

S4: Shifting current expenses to a later period

S5: Failing to record or disclose liabilities

S6: Shifting current income to a later period

S7: Shifting future expenses to an earlier period

GET IT "RIGHT" THIS TIME

Justifying a claim that one's product is superior to a competitor's is often not a simple matter. If one approach doesn't work, it may mean that there is no significant difference in the products or that the approach taken was not sophisticated enough to reveal the expected difference. The idea suggested is that another approach of greater sophistication be employed to test for a difference. There is no question about the propriety of this approach.

However, when big money is at stake there is temptation, sometimes overwhelming, to employ a fudge factor to stack the deck in one's favor. Sterling Drug conducted a statistical analysis which showed no difference between Cope and aspirin. The researchers then employed a different statistical method to obtain the results expected and desired. Based on the testimony of a witness the Federal Trade Commission found credible, it concluded that the second test was "a gross and obvious example of statistical manipulation, and . . . not acceptable scientific methodology" [8].

CONFLICTS OF INTEREST

The preceding case illustrates one way in which the data deck might be stacked to favor a preferred outcome. There are a number of ways in which this might happen and, in the spirit of "to be forewarned is to be forearmed," we look at some of them here.

Data Suppression by Contract. In 1987 the predecessor of the Knoll Pharmaceutical Company hired Dr. Betty Dong of the University of California at San Francisco to do research on its thyroid drug Synthroid. Their contract gave Knoll's predecessor veto rights over any publication. The research, completed in 1990, found that Synthroid was no more effective than less expensive generic versions of the drug, which was not what Knoll wanted to hear. They exercised their veto right and only 7 years later was Dong allowed to publish her findings (*Journal of the American Medical Association*, April 16, 1997).

If You Don't Want to Know, Don't Ask. In the late 1970s the Philip Morris Company set up a highly secret research unit to study nicotine and its effects on the body. Scientists hired to work in the unit had to sign an agreement which bound them not to discuss or publish their research without Philip Morris's permission. Two key researchers, Drs. Victor DeNoble and Paul Mele, who worked in the unit from 1980 until it was closed in 1984, painted a startling picture of a hush-hush operation and data suppression in testimony given before a congressional hearing on April 28, 1994.

Among the findings suppressed were data indicating that nicotine was addictive to laboratory rats; that another common product of tobacco combustion, acetaldehyde, was addictive; that a combination of nicotine and acetaldehyde sharply increases the addictive power of each. Also suppressed were the discovery of a nicotine-like substance, called 2′ (read two prime) methylnicotine, which causes animals to experience the same "high" as nicotine, but without toxic effects on the heart, and the finding that secondhand smoke has a toxic effect on plants.

Although the findings were preliminary, they were excited about where they might lead, the scientists stated, but then it all came to an abrupt end. On April 5, 1984 they were told to halt their studies, kill the laboratory animals, and turn in their security badges by morning, the scientists testified.

No, It's Not Politically Correct. In 1972 Dr. Carl Seltzer sought to publish a paper showing that men who consumed moderate amounts of alcohol had a lower risk of heart disease than those who did not drink. His work was supported by the National Institutes of Health, which refused to grant him permission to submit the paper for publication, arguing that "an article which openly invites the encouragement of undertaking drinking with the implication of prevention of coronary heart disease would be scientifically misleading and socially undesirable in view of the major health problem of alcoholism that already exists in the country" [9]. Since that time other studies have shown a positive correlation between consumption of moderate amounts of alcohol and lower risk of heart disease.

Velvet Glove, Iron Hand. "There are so many ways to influence what works get published," commented Dr. Alan Garber, associate professor of medicine at Stanford University, "including how a drug company funds research, the grants they give, whether those writing articles that are favorable to a particular drug or device get on a speakers list, get honoraria, do consulting" [10]. From a scientist's perspective, there are certain advantages to not making waves.

The Other Side of the Coin. Does a scientist's financial interests, such as stock ownership, consultancies, and patent rights, introduce bias into his or her work? The answer, of course, is not necessarily. Should the reader of a research article be aware of potential financial conflicts of interest of the article's author? According to a study by Sheldon Krimsky of Tufts University and L. S. Rothenberg of the University of California at Los Angeles, 0.5 percent of some 62,000 articles published in 1997 included information on the authors' research-related financial ties [11].

In a related situation involving doctors, throughout much of the 1980s doctors could refer patients to treatment centers in which they had a financial stake. The practice was outlawed when it was found that doctors were overusing treatments and tests when they had financial interests in the centers that provided them [12].

Informative, but chilling, is the case of interventional cardiologists discussed in "When Physicians Double as Entrepreneurs" [13] and biomedicine scientists as entrepreneurs [14].

WOUNDED MILITARY FIGURES

In the spring of 480 B.C. the Persian King Xerxes launched against Greece the largest military operation the world had seen to that point. But how large? The Greek historian Herodotus, with tongue in cheek, reports the size of the fighting force as 2,641,610 men. After considering the service train that went with them, Herodotus put the total force at 5,283,220 men. But, he adds:

No one could give the exact number of women who baked the bread, or of the concubines, or the eunuchs, or the transport animals and baggage-carrying cattle and Indian dogs that came with the army—of all these creatures no one could count the numbers, they were so large [15].

The invasion force was large, no doubt, but 5,283,220 men plus others? This exceeds by far the entire population of Greece at that time [16].

Moving across a span of 2500 years to modern times we find a reaffirmation of the observation that the first causality of war is the truth. The NATO bombing of Kosovo in 1999 brought claims of destroyed Serbian military targets, as noted in the second column of Table 2.1. The number

Table 2.1 Claimed versus Found

	Number NATO claimed were destroyed	Number found in Kosovo
Tanks; self-propelled artillery	110	26
Armored personnel carriers	210	12
Mortars; towed artillery	449	8

Source: Ref. 17.

subsequently found to have been destroyed is given in the third column of Table 2.1.

The lesson is clear: He who takes military statistics at face value is a potential buyer for a soup sandwich.

A QUESTION OF "ACCURACY"

Many people believe that the accuracy of a numerical result is increased by expressing the value in terms of a large number of decimal places. That this is not necessarily the case is illustrated by the following experiment. Take a tub of water and place 7 empty pails on the floor. Spill the water from the tub into the pails, giving each an equal share, as nearly as possible. It would be appropriate to say that each pail contains roughly one-seventh of the tub of water. While this may not seem very accurate, it's the best that can be said. If, however, we divide 1 by 7 to obtain 0.142857 . . . , and proclaim that each pail contains about 14.2857 percent of the water of the tub, one might be deluded into believing that this is a more accurate description of the amount of water in each pail if one equates accuracy with number of decimal places. From this point of view the accuracy of the description would be further enhanced by carrying the division of 1 by 7 to still more places, 0.1428571428571 . . . , for example. It's nonsense, of course; we are no better off than when we said that-

each pail contains about one-seventh of the tub of water.* The important point is: Real accuracy depends on the accuracy of the data.

In calculating with approximate numbers, we should keep in mind that the accuracy of a result cannot exceed the accuracy of the least accurate number in the data. The sum of the approximate numbers 2.34 (known to two places) and 1.131 (known to three places), for example, should be recorded as 3.47 (to two places) rather than 3.471. The number 3.471 would give a misleading sense of accuracy.

* M. Richardson, *Fundamentals of Mathematics*, rev. ed. (New York: Macmillan, 1958), p. 402.

3

Coming Up with the Numbers

It is often clear to the mind's eye what data are required. But the mind's eye does not have to contend with reality, which makes the task of obtaining the data more easily contemplated than accomplished.

FOR "REAL" ACCURACY, COUNT THEM ALL. RIGHT?

A sample provides us with a view of the population from which it is drawn, an excellent view if it is "properly" extracted. But would it not be preferable to look at the entire population? The short answer is that it depends; the long answer involves consideration of what it depends on. What it depends on includes the nature and size of the population we seek to get a grip on, degree of accuracy required, and cost.

To count the number of defectives coming off a production line, a randomly chosen batch is inspected, the number of defectives in the batch is determined, and an extrapolation is made to the production line as a whole. This procedure gives sufficient accuracy, is cost effective, and timely. If there are good indications that the number of defectives is out of control, the production manager wants to know about it as quickly as possible so that the production process can be adjusted. Sampling accommodates this constraint. An audit of a company doing a high volume of business is accomplished by examining a randomly chosen sample of transactions and extrapolating to the whole. Examining every transaction could take an inordinate amount of time and resources with less accurate results.

To count the number of fish in a lake, it is hardly feasible to do a direct count by catching them all and keeping track of how many were pulled in. In this situation, and others like it, a technique called capture-mark-recapture is employed. A large number of fish are caught, tagged, and released back into the lake. A reasonable amount of time is allowed to elapse to allow the fish to disperse (a few days, perhaps), another batch of fish is caught, and note is taken of how many were previously tagged. If 1 percent of the second batch were previously caught, the number of fish caught in the first sample is taken as an estimate, called the maximum likelihood estimate, of 1 percent of the number of fish in the lake, which gives us an estimate of the lake's fish population. If 200 fish, let us say, were caught in the first batch, 200 would be our estimate of 1 percent of the fish population, which yields a maximum likelihood estimate of 20,000 for the fish population.

POLLING AND ITS PERILS

Counting fish in a lake is one order of business, but how about people? How do we obtain a grip on their preferences, attitudes, and beliefs? Since it is often not feasible to reach all, the idea suggested is that we choose a sample from the population of interest to us, administer a questionnaire (called a survey) to the members of the sample chosen, and use the results as a reflection of the beliefs of the population. This practice is called polling, and the question that immediately arises is, how accurate are polls? The answer depends on a number of factors.

The 1936 presidential election was on the horizon with Franklin D. Roosevelt, seeking reelection, facing the Republican nominee Alfred E. Landon and third party candidates. Riding high from past polling successes, *The Literary Digest*, a popular weekly magazine of the time, geared up. It mailed some 10 million ballots to prospective voters and eagerly awaited the returns.

Against this background it seemed foolhardy in the extreme for George H. Gallup to take on *The Literary Digest*. Gallup founded the American Institute of Public Opinion in 1935 and initiated a weekly column called "America Speaks." To attract newspaper subscriptions, he offered a money-back guarantee that his prediction of the presidential winner would be more accurate than that of the *Digest*. Gallup predicted a Roosevelt victory with about 54 percent of the popular vote. In response to its poll, the *Digest* received 2,376,523 responses, with 1,293,669 for Landon, 972,897 for Roosevelt (42.9 percent of the two-party vote) and the remainder for third party candidates. The *Digest* predicted that Landon would carry 32 states with 370 electoral votes and that Roosevelt would carry 16 states with 161 electoral votes.

The actual vote gave Roosevelt a landslide victory with 62.5% of the two-party popular vote. He carried 46 instead of 16 states and received 523 instead of 161 electoral votes. Already in financial trouble, *The Literary Digest* suffered a devastating blow to its credibility and folded the following year.

Independently of Gallup, two other researchers, Elmo Roper and Archibald Crossley, using similar methods, also predicted a Roosevelt landslide.

Lessons

The 1936 Roosevelt landslide carried with it a number of lessons for the practice of polling in general and election polling in particular.

Sample Size. The accuracy of a poll is not determined by enormous sample sizes, impressive as they may appear to be. The major point is that the sample must be "properly" chosen. Preelection poll samples these days tend to be between 1000 and 2000 in size.

Target Population vs. Sampled Population. It is most important that the population actually being sampled be the one about which we seek to draw inferences. The failure of *The Literary Digest*'s 1936 poll was in part due to its failure to reach the target population of all prospective voters. To a large extent the mailing list employed by the *Digest* came from telephone directories and automobile registrations, which was biased in favor of wealthier Americans who tended to vote Republican.

Nonresponse Error. Of 10 million odd ballots sent out by the *Digest*, 2,376,523 returns were received, for a 24 percent response rate. The other side of this coin has a 76 percent nonresponse rate. What were the nonrespondents thinking? It's hazardous to predict one way or another. Those who did respond wanted to make sure that their opinion was counted. Those who did not respond did not care enough about the poll to respond.

In some locations, such as Allentown, Pennsylvania, the sampling list was not drawn from telephone and automobile registration lists, but from voter registration lists with no inherent Republican bias. The poll's results show a clear nonresponse bias in that the proportion who favored Landon in the poll was much higher than the proportion who favored him in the election. Those voters who came from the lower economic *strata* tended not to respond to the *Digest*'s poll when invited to do so and they strongly favored Roosevelt.

Further Lessons: Dewey vs. Truman. Other lessons were to come, but the laboratory of reality required more time to make clear that additional fine-tuning was needed. The Dewey vs. Truman election of 1948 provided some of these lessons. Vice president Harry S. Truman became president on the death of President Franklin D. Roosevelt on April 12, 1945. It was a difficult post-war period and when Truman was nominated by the Democratic Party as its presidential candidate in 1948, he faced an uphill fight against the Republican challenger Thomas E. Dewey, governor of New York. The polls showed Dewey comfortably ahead and he adopted a strategy of caution and platitudes ("Our future lies before us"), seeking to avoid making commitments and enemies. The political establishment,

with the exception of Truman, felt that Dewey had it in the bag. Truman vigorously counterattacked, delivering 300 odd "give'em hell" speeches, as they were called, on a train tour of the country. Truman went to bed early election day evening and woke up to find himself president for another term. The pollsters and political pundits had missed the boat.

Stability of Voter Opinion. George Gallup's last preelection poll, taken between October 15 and 25, showed Truman trailing Dewey by 5 percentage points, 49 to 44 percent. Almost all pollsters believed that public opinion was pretty well set by this time and would change little in the time remaining before the election. At this point they closed down their polling operations. Reality, of course, proved them wrong.

Undecided Vote. How is the undecided vote to be handled? Gallup split the undecided vote in the same proportion as those who had expressed a preference for Dewey and for Truman, which strongly favored Dewey. Subsequent analysis showed that 14 percent of the undecideds made up their minds in the last two weeks of the campaign and that 74 percent of these went for Truman. It is clearly hazardous to decide for the undecided.

Sampling Methods. The major objective of sampling is to choose a sample which, in some sense, provides a good reflection of the attitudes of the larger population from which it is taken. With one sampling method, called quota sampling, an interviewer is instructed to interview an assigned number, or quota, of individuals in groups defined by specified characteristics such as race, religion, ethnic background, age, occupation, and economic status. Which individuals are chosen to be interviewed is left to the discretion of the interviewer, as long as the specified quota is met.

With random sampling, on the other hand, it is envisioned that the sample is chosen in such a way that there is no bias, deliberate or inadvertent, which favors certain samples being chosen over others. Random sampling is an equal opportunity procedure, whereas the way that a quota is satisfied in quota sampling is dependent on the judgment and convenience of the sample taker.

If it is desired that the sample be chosen in a way that group characteristics such as race, ethnic background, and so on, are reflected in the sample, but that the sampling within these groups be unbiased, then one may employ stratified random sampling. This method envisions the population from which the sample is to be drawn as consisting of nonoverlapping components (such as the states of the United States, for example), called *strata*, where a part of the total sample is chosen at random from each *stratum*.

The quota sampling method employed by Gallup, Roper, and Crossley in the 1936 election was a vast improvement over the mail poll conducted by *The Literary Digest* in that it yielded results more representative of the population at large and gave much better control over nonresponse error. Gallup was much closer to the mark than the *Digest* in 1936, but his predictions were wrong on six states and he predicted a Roosevelt victory with about 54 percent of the popular vote, 7 percentage points off the mark. It might not matter when the outcome is strongly one-sided, but in a close election it could make a significant difference. The Dewey vs. Truman contest provided a decisive test case. It was a close election; in electoral vote terms, Dewey would have won had he carried California, Illinois, and Ohio, each of which he lost by less than 1 percent. The size of Gallup's and Crossley's error in predicting Dewey's victory in 1948 (5.4 and 4.7 percentage points, respectively) was smaller than the error in predicting Roosevelt's victory in 1936 (7 percentage points for each). But in terms of the bottom line, wrong is wrong.

Polling's failure in 1948 brought matters to a head concerning the sampling method to be used. Random sampling was the technique favored by many critics, but it was more complicated and much more expensive to implement than quota sampling. Gallup had serious doubts that the increased accuracy achieved would be worth the expense, but after the 1948 election he and other pollsters switched to random sampling for preelection polls.

BEWARE OF . . .

Questions

Nuances in Wording. The wording of poll questions is, with the best of intentions, a delicate matter. In an experiment conducted by Elmo Roper as far back as 1940, of those asked if the United States should forbid public speeches against democracy, 46 percent replied "no." Of those asked if the United States should allow public speeches against democracy, 25 percent said "yes." Support for free speech was much greater when the term "forbid" was used rather than "allow."

In the fall of 1997 the following question appeared on a referendum ballot in Houston. "Shall the charter of the City of Houston be amended to end the use of affirmative action for women and minorities?" The initiative was defeated by a 55 to 45 percent margin. But is its wording equivalent to that taken from the 1964 Civil Rights Act? "The City of Houston shall not discriminate against, or grant preferential treatment to, any individual or group on the basis of race, sex, ethnicity, or national origin." Judge Sharolyn Wood argued that it was unfair to change the original wording, threw out the results of the Houston referendum, and ordered a new vote.

Meaning of Words. "Do you think President Nixon should be impeached and compelled to leave the presidency, or not?" In a Gallup poll held in July 1974, 24 percent of the respondents said "yes." When the question was posed as "[Do you] think there is enough evidence of possible wrong doing in the case of President Nixon to bring him to trial before the Senate, or not?" 51 percent said "yes." This was in response to a Gallup poll held at the same time. A case of fickle respondents? No. The second wording makes clear that Nixon would be brought to trial, which was not as clear to respondents from the first wording because the meaning of "impeached" was in doubt.

Questions involving terms which might be understood one way in everyday usage, but have a specific technical meaning, must be handled carefully. Robbery, involving confrontation between victim and offender, for example, is technically different from burglary, which does not involve personal confrontation. Such a distinction must be made clear in any question involving such terms.

Poorly Formulated Questions. Poorly formulated questions can, not unexpectedly, bring an ambiguous response. This question was asked in a 1992 poll conducted by the Roper Organization for the American Jewish Committee: "As you know, the term 'Holocaust' usually refers to the killing of millions of Jews in Nazi death camps during World War II. Does it seem possible or does it seem impossible to you that the Nazi extermination of the Jews never happened?" It's not unreasonable to expect the double negative ("seem impossible . . . never happened") to cause confusion, and this is what the responses received reflect.

A study released by the Education Department in September of 1993 concluded that half of the adults in the United States cannot read or handle arithmetic. This is certainly an alarming figure, but is it accurate? That, to a large extent, depends on the success of the test makers in developing questions free of cultural bias, verbal ambiguity, and distracting irrelevancies.

The following arithmetic question was posed:

> The price of one ticket and bus for *Sleuth* costs how much less than the price of one ticket and bus for *On the Town*? A charter bus will leave from the bus stop (near the Conference Center) at 4 p.m., giving you plenty of time for dinner in New York. Return trip will start from West 45th Street directly following the plays. Both theaters are on West 45th Street. Allow $1^1/2$ hours for the return trip. Time: 4 p.m., Saturday, November 20. Price: *On the Town*, ticket and bus: \$11.00; *Sleuth*, ticket and bus: \$8.50. Limit: Two tickets per person.

The question itself raises the question of to what extent it is intended to test arithmetic and to what extent it is intended to test one's ability to successfully negotiate a verbal maze, particularly when many of those participating in the test are foreign-born adults whose first language is not English. With tests of this sort, the underlying question of how well the questions achieve their intended objective must be given careful consideration before meaningful conclusions can be drawn about the population being studied.

Long and/or Loaded. Greg Schneiders and Jo Ellen Livingston note that Microsoft claims that the results of a survey it commissioned indicate overwhelming support for incorporating its Internet browser into the Windows platform. "The 350-word question," they point out, "mostly argued in favor of incorporating the browser; nothing in the question argued against it. What's remarkable is that 15 percent of respondents could actually bring themselves to disagree" [1].

Direct Questions. Direct questions may provide a misleading response. Do you intend to vote in the forthcoming election? Many people would answer yes rather than run the risk of being thought an irresponsible citizen.

Personal Questions. Personal questions concerned with drug use, one's sexual behavior or preferences, for example, may provoke a response considered socially acceptable rather than candid.

Question Order. Subtle differences in question ordering can have a significant impact on the responses obtained. The January 1984 pre-presiden-

tial election *New York Times* / CBS poll found that voters preferred incumbent President Ronald Reagan to Democratic challenger Walter Mondale by 16 percentage points; the question was posted at the beginning of the interview, which favored the better-known Reagan. Gallup and *Washington Post* / ABC News polls taken around the same time posed the question near the end of the interview after the questions about Reagan's policies had been asked; this helped Mondale because he had less of a record to defend. These polls had Reagan and Mondale about even.

Response Options

The two-category response option of the form Yes or No is much more restrictive than the four category response option Yes, No, Not sure, Not enough information. The number of response options available may profoundly influence the response given, including the possibility of nonresponse. It is interesting to compare voter reactions in two July 1992 opinion polls on President George Bush and Democratic party candidate Bill Clinton. The responses for the two polls are shown in Tables 3.1 and 3.2.

Table 3.1 Three Response Options

	Bush	Clinton
Favorable	40%	63%
Unfavorable	53%	25%
Don't know	7%	12%

Table 3.2 Five Response Options

	Bush	Clinton
Favorable	27%	36%
Unfavorable	49%	24%
Undecided	22%	31%
Haven't heard enough	2%	9%
No answer	1%	1%

Pseudo-Opinions

On one end of the spectrum nonrespondents affect the reliability of a poll. At the other end there are those who are eager to respond, but who literally don't know what they are talking about. In an experiment conducted some years ago a group of researchers at the University of Cincinnati asked a randomly chosen group of people whether the 1975 Public Affairs Act should be repealed. About half of them expressed an opinion on this nonexistent act.

The Effect of Being Polled

Being polled itself may put people on guard if they feel they might lose benefits or be penalized in some way if they give the "wrong" answers. In the 1930s many people on welfare were afraid they would be thrown off the welfare rolls if they gave an undesirable response. The election of Violeta Chamorro as president of Nicaragua in early 1990 was contrary to poll predictions which had Daniel Ortega with a substantial lead. One reason the polls were so inaccurate was that the intimidation factor was not accurately taken into account. Nicaragua had been under authoritarian rule and in a state of civil war for a number of years, and many voters were not about to freely express their political preferences to pollsters. In such situations it is especially important to go all out to win the trust of respondents.

Interviewer-Induced Bias

Conducting a "successful" interview, whether in-person or by telephone, is not a simple matter. People respond to interviewers as well as questions and the interviewer must strike a balance in being personable, respectful, and considerate of the person being interviewed, and professional. The type of person who sets your teeth on edge by the way he or she says "Good Morning" is not likely to be a successful interviewer.

It was recently found that computers elicit more honest responses to delicate personal questions than human beings [2,3]. The more socially stigmatized the behavior, the greater the difference in response, it was found. This finding threatens to call into question data that had previously been obtained on a number of sensitive subjects.

The Final Gap

"There's many a slip between the cup and the lip," the old saying goes and this is especially the case with the gap between the last poll taken and an election itself. A poll may be likened to a picture taken by a camera at a point in time. Looking at a sequence of polls is analogous to looking at a sequence of still photos, a videotape if you will. While this might be strongly indicative of what will happen, it is not the same as what does happen.

THE SCOPE OF POLLING

Today the practice of conducting polls, or surveys, as they are also called, is widespread with an enormous range. Political preference polls are perhaps the best known, but the majority of surveys that are conducted are concerned with specific social, commercial, or administrative concerns that do not occupy the public's spotlight with the same intensity. Government agencies conduct surveys to obtain information about employment, who uses food stamps, and who uses the national parks. TV networks employ surveys to obtain information about the popularity of their programming; businesses employ surveys to determine consumer preferences and reactions to their products; and interest groups conduct surveys to obtain a sense of the public's reaction to such matters as the state of education, the economy, crime, race relations, health costs, environmental issues, gun control, foreign policy issues, and the war against drugs.

The problems inherent in carrying out a poll, or survey, to obtain reliable data and project them on a larger population are clearly considerable, which should give us pause to closely watch the trouble spots and lead us to be cautious in interpreting the results obtained.

IS THERE MORE TO IT THAN WHAT THE POLLS SAY?

"It's much safer to stay here in Washington and rely on the polls for your readings for what the people out there actually are thinking. Do that and you will remain convinced that Americans are bored to tears with Clinton scandals and almost completely absorbed in enjoying their own lives," observed veteran political columnist Godfrey Sperling in an article published in June of 1998 [4]. But Sperling ventured out, once to the West Coast and then to the Midwest. "During that time I listened in on or was part of a number of conversations. And I also asked questions of many people. . . . my reading, unscientific of course, but at least gained from first-hand reporting, is that the polls are wrong. . . . I've now found a public that may be telling pollsters this while really keeping a close eye on and intense interest in the Washington Scene. . . . Yes, whether I'm entitled to or not I've become convinced that there's no new lasting public tolerance emerging—one that will put up with anything this president is accused of doing, even if there is strong indication that he has done it. There is latent outrage out there, among those who keep saying, over and over again, how bored they are with what's going on in Washington."

Political scientists Lawrence Jacobs of the University of Minnesota and Robert Shapiro of Columbia University have strongly criticized the way that polling on public policy and the follow-up reporting has come to resemble election polling. 'The media's use of polls is producing shallow and myopic reports that track the political strategy of insiders rather than conveying substantive information on the public's preferences toward entitlements. Superficial and nonsubstantive reporting has become a staple of the fluid world of electoral politics, where candidates often practice the art of ambiguity. It is striking, through, to see a familiar pattern of shallow reporting on the weighty topic of Social Security, Medicare and entitlements in general' [5].

MEDIA COVERAGE OF POLLING

The heart of the matter centers on determining "appropriate" data which is reliable about the public's attitudes and presenting it to the public in a timely manner. Polling organizations are faced by the first problem and the history of polling tells us that polling professionals have seriously sought to improve polling and survey research. The "Beware of . . ." section of this chapter is based on this ongoing search for improvement.

Presenting polling results to the public in a reliable and timely manner is the important second dimension, which is the responsibility of the media. How good a job have the media done? "The problem isn't polls themselves," observe Lawrence Jacobs of the University of Minnesota and Robert Shapiro of Columbia [6]. "Understood properly, polls in the 1996 presidential election have generally provided reasonably consistent results. The problem is that journalists' current coverage of campaign polls is seriously flawed. Media reports on polling are probably doing more harm than good. Americans are being deceived by some of the polling information presented to them. Something must be done" [7].

Problems

Horse Race Emphasis. Fifty-two percent of the newspaper stories Jacobs and Shapiro reviewed only mentioned polls in connection with who was ahead. More than 7 out of 10 stories devoted some part of their report to harping on polls that sized up Clinton's lead. Eighty-five percent of the television stories reviewed gave attention to the horse race dimension.

Attention to Character and Issues. Thirteen percent of the newspaper stories they reviewed focused on polls on candidate character, 12 percent on issues coverage, 23 percent mixed poll results on the horse race dimension with other issues.

Something Out of Very Little? While a poll which has A leading B by 52 to 48 percent, with a sampling error of 5 percent, let us say, is statistically indistinguishable from a similar poll taken at roughly the same time which has A leading B by 53 to 47 percent or 54 to 46 percent with a sampling error of 5 percent, the tendency in media coverage was to ignore the sampling error dimension and hype the result as indicating movement in the electorate or electorate volatility.

Home Team Polls. The tendency of the major news media was to focus on the polls they sponsored to the neglect of others.

Premature Projections. Relying on a New Hampshire exit poll of voters by the Voter News Service, Dan Rather of CBS News reported at 7:00 p.m. of election day 1996 that Democratic challenger Dick Swett had defeated incumbent Republican Senator Bob Smith. Hoping to score a news coup, CBS reported a projection as fact. As it turned out, Senator Smith retained his seat. CBS established that the lesson of the classic UPI photo showing President Truman holding up on early edition of the *Chicago Daily Tribune* with the headline "Dewey Defeats Truman" had not been fully grasped. There was a lesson here too for the pollsters concerning the reliability of exit polls.

SEXUALITY BY THE NUMBERS, OR NOT?

In 1987 Shere Hite published *Women and Love: A Cultural Revolution,* her third book on human sexuality. In her first two books, *The Hite Report* (1976) and *The Hite Report on Male Sexuality* (1981), Hite restricted herself to telling what she had learned from women and men who replied to the extensive questionnaires concerning their sexual problems and attitudes she had circulated. For her first book she circulated approximately 100,000 questionnaires, from which she received 3019 responses for a response rate of about 3 percent. Four different versions of the questionnaire were sent to women's organizations that were asked to circulate them. A similar methodology was employed for her second book. Approximately 119,000 questionnaires were distributed, with a response rate just under 6 percent being obtained. Shades of *The Literary Digest*'s debacle come to mind, but Hite was not claiming that her sample was representative of women and men in general. Hers was a qualitative rather than statistical study.

In statistical studies, the same questions must be asked of all prospective respondents with the same response options being available to them all. Uniformity of the underlying conditions and the choosing of a "representative" sample so that the results obtained could be projected onto the population at large are essential for a statistical study. Qualitative studies, on the other hand, focus on the special qualities of each individual potential respondent. Capturing the diversity inherent in individuals takes priority over ensuring uniform underlying conditions. It is not a matter of one kind of study being superior to the other, but rather of which methodology is appropriate to the study being undertaken. Although her first two books raised much controversy, Hite was on safe methodological ground.

In her third book, Hite attempted to cross the bridge from the qualitative results she had obtained to statistical generalizations about sexual attitudes of women in America. The bridge collapsed. Her methodology was almost universally criticized. ABC News in conjunction with *The Washington Post* conducted a telephone poll for October 15–19, 1987 to see if they could duplicate her results. They could not; their results were sharply at variance with those projected by Hite [8]. To take two examples, she found 84 percent of women as not being satisfied emotionally with their relationships; ABC/WP found 7 percent of married women and single women in a relationship as not being emotionally satisfied. Hite found 78 percent of women feeling they are only occasionally treated as equals most of the time; ABC/WP found 81 percent of married women and single women in a relationship feeling they are treated as equals most of the time. There were differences in the way questions were posed, but they were not startling. As to who was closer to the mark, ABC/WP clearly takes the Trustworthy Prize because of its sound statistical methodology.

THE FUTURE OF POLLING?

An on-line survey conducted during the summer of 1999 by ABC.com invited people on its website to state their preference for president in a hypothetical Democratic primary consisting of Vice President Al Gore, former New Jersey Senator Bill Bradley, and actor Warren Beatty. Beatty received about 25 percent of the vote, Gore 45 percent, and Bradley 29 percent.

Should such a result be taken seriously or should a warning be posted: Pseudo-poll for entertainment only? The latter, of course, because on-line polls rely on self-selected respondents and violate all principles of sound polling that have emerged since *The Literary Digest*'s disastrous Landon-Roosevelt presidential poll of 1936.

Flawed as they may be, on-line polls have grown in popularity. Such is the hypnotic pull of numbers to seize our attention that reporters and commentators affiliated with reputable news organizations have begun to cite, without warning, the results of on-line polls in their commentaries. This serves to confer on them a false sense of respectability. Beware.

The good news is that polling organizations have sought to strengthen the reliability of on-line polls by combining the best of the new and avoiding the worst of the old. The basic idea is to draw random samples from the target population, contact the individuals by phone, and then hook them up to the internet. This establishes a randomly selected subpopulation whose members could be queried over the internet.

How well has it worked? Humphrey Taylor, chairman of the Harris poll, and colleagues, argue that: "The November 2000 elections provided Harris Interactive with a unique opportunity to test the accuracy of our on-line survey methods and, specifically, our ability to predict elections. Happily, the accuracy of our efforts in the 73 different races we covered exceeded our most optimistic expectations" [9]. Details in support of this view are provided. They also express the view that "the ability to conduct reliable online research will depend on a continuing investment in testing and improving different weighing schemes" [9].

It sounds promising, but . . .

4

Political and Ideological Coloring

CENSUS POLITICS

As we observed in *Polling and Its Perils*, "Counting fish in a lake is one order of business, but how about people?"

In seeking to determine the population of the United States and its characteristics in the mandated decennial census, the approach used in the census of 1990 was similar in many respects to that used in the first census taken two centuries ago in 1790, direct count. It is based on the view that people are in a particular location at a particular time and that it is basically a matter of having them fill out the required questionnaire and then tabulating the results. In theory straightforward, in practice not so. The homeless, migrant workers, and people on the move are difficult or impossible to locate; the questionnaire is unclear to many who are located and asked to fill it out; those who refuse to fill out or return the census questionnaire are not insignificant in number (in some large cities response rates fell under 50 percent in the 1990 census); mispunching errors create whole categories of nonexistent people; some population subgroups are overcounted, while others are undercounted, and on it goes. The 1990 census found 248.7 million people living in the United States. A post-census analysis estimated 4 to 6 million people as not having been counted. Those missed in the count were, for the most part, minorities and urban poor who tended to vote Democratic.

In a legal, if not actual numerical, sense we have an "exact" count. Alas, the traditional direct-head-count method, as it is usually referred to, does not yield the accurate count in practice that one might believe it capable of in theory.*

For the 2000 census the stakes promised to be high. The population count would determine how an estimated $180 billion a year of federal aid is distributed, congressional seats are apportioned, congressional districts within states are determined, and state legislative boundaries are set. The method employed for carrying out the census goes further than being of arcane theoretical interest. A great deal of money and political power are at stake. That the use of sampling tools can produce a more accurate census is widely accepted in statistical and mathematical circles. The Clinton administration proposed use of such a tool to correct for the undercount of minorities and urban poor that occurred in the 1990 census. Apart from questions of what a fair count and legal count entail, it is, for the most part, in the political interest of most Democrats to support this proposal. Most Republicans have insisted that the traditional direct-head-count method (which itself involves a degree of statistical sampling) be used, but that more census workers be employed and that their efforts be intensified.

Prompted by lawsuits, the issue came before the Supreme Court, which in January of 1999 rejected the Clinton administration's plan to conduct the 2000 census with the support of statistical sampling, ruling by 5 to 4 vote that the Census Act made mandatory use of the traditional direct-head-count method for obtaining the figures for apportioning seats in the House of Representatives. The Court also indicated that statistical adjustments might be made for other purposes of the census, including the setting of congressional boundaries within states and the distribution of federal funds.

For further discussion of census politics, see Ref. 1.

* The term "traditional direct-head-count method" evokes a sense of accuracy born of simpler times in which census takers would line up everyone in a community and count them, one by one. This never happened. The census taker relied on the information provided by whomever in a household chose to provide it, erroneous or not; verification was not attempted. If a census taker was not able to locate a household, he would rely on information provided by whomever he could find that might know—landlords, neighbors, for example.

SOVIET MILITARY OUTLAYS: REALITY OR INVENTION?

During the confirmation hearings held in the fall of 1991 on the fitness of Robert Gates to be CIA director, senior CIA analyst Marvin Goodman and two other CIA analysts testified that while serving as deputy director of intelligence Gates tailored intelligence estimates to suit administration policy in several important areas. One of the areas cited was concerned with Soviet capabilities and intentions.

In connection with this state of affairs a leading American analyst of Soviet military spending, Franklyn D. Holzman of Tufts University comments [2].

Professor Goodman's revelations come as no surprise to me; indeed, they suggest an answer to puzzles raised over the last 15 years in my studies of CIA estimates of Soviet military spending.

The CIA estimates have been, in my opinion, continually slanted upward to make Soviet military spending appear larger than careful analysis of the data suggests, thereby supporting the Reagan-Bush-Pentagon military buildup policies.

It was clear to me that the decisions to slant the conclusions were made at the highest levels because the techniques used were so unacceptable on scholarly grounds. In my work on military spending, I have come to know many of the CIA analysts who prepare these estimates and I believe that, had they been free of political pressure, the estimates would not have been distorted.

As to the implications of these developments for the post cold war military budget, Holzman argues [3; see also Refs. 4–6]:

> Had there not been a CIA error followed by a cover-up, our military spending might have continued to increase at a slower rate. It is estimated that for the decade 1979–1988 the United States would have spent on defense approximately $800 billion less in present-day prices. . . . The Pentagon used exaggerated estimates of Soviet military spending to help get enormous budget increases. Now that the cold war is over, this overspending should be taken into account in evaluating the Pentagon's claims to severely stretched resources.

FUDGE FACTORIES

The Slippery Number Society has an international clientele. A few examples follow.

Brazil

In a frank conversation between television interviews that was inadvertently broadcasted across his country, Brazil's finance minister Rubens Ricupero expressed the sentiments of many kindred spirits when he confessed of economic indicators: "I have no scruples, what is good we take advantage of. What is bad, we hide" [7]. Minister Ricupero was immediately dismissed, but was this because of his performance or indiscretion?

China

Chinese government statistics have run a gamut of slipperiness. After the Communist party assumed control in 1949, government statistics were systematically distorted to serve the wishes of the new political establishment. During the period of the Cultural Revolution of the late 1960s and early 1970s, data gathering was abandoned as unscientific.

Since the passing of the Cultural Revolution, data gathering and the publication of state statistics have resumed and other pressures have developed. In May 1994 Zhang Sai, director of the State Statistical Bureau, "warned that distorted statistics are increasing tensions between Beijing and localities" [8].

Foreign investors in China are wary of Chinese statistics and many have taken to generating their own.

Japan

By late July 1998 American financial experts reached the conclusion that the magnitude of Japan's banking crisis was far worse than had been publicly acknowledged. The bad debts were estimated as being on the order of $1 trillion, nearly twice the official estimate. The true amount, financial experts emphasized, was hard to pin down because Japanese banks had been using accounting tricks to conceal debts that are not being paid [9]. Sound familiar?

On May 25, 2000 the Japanese government announced that it would correct a deliberate omission of data concerning the gross domestic product (GDP) for October–December 1999. Inclusion of the data would show that GDP in this quarter declined by 1.6 percent from the previous year; its exclusion had GDP declining by 1.43 percent, which would make it less difficult for the government to reach its target of 0.6 percent economic growth in the fiscal year that ended March 31. Although the government faced an election in a month, the head of the Economic Research Institute of the Economic Planning Agency insisted that the omission of data was not done for political reasons [10,11]. But Kenji Okue, an economist with the investment bank Dresdner Kleinwort Benson, noted: "In Japan, everything, including economic statistics, is about politics" [10].

Soviet Union and Russia

From the beginnings of the Soviet state, Soviet statistics acquired a reputation of being unreliable [12,13]. Writing in 1990, V. N. Kirichenko, chairman of the USSR State Committee on Statistics, expressed a hope to "ensure the accuracy of the data . . . restore the trust in such data on the part of the Soviet and international public. The country can no longer afford to seek the right way with the help of trick mirrors" [14, 15].

Since the breakup of the Soviet Union, Russia has continued to have problems with government statistics, but for a different reason. Rather than exaggerating output, the statistical pendulum had swung to the extreme of underestimating it. In June 1998 Russia's top statisticians were arrested on charges of manipulating data to underestimate the production of Russian businesses to help them minimize their tax obligations [16].

United States

Budget Fudge. As David R. Francis, senior economics columnist for *The Christian Science Monitor*, put it: "It's a good thing the SEC [Securities and Exchange Commission] doesn't turn its stern eyes on its political masters—the White House and Congress—who play accounting games to please (or deceive) the voting public. If it did, some politicians would probably end up in the hoosegow" [17]. For some specific examples it would be best to consult Francis's article directly. Also see Refs. 18–23.

Give Us the "Right" Numbers. In June 1998 the thrust of the Republican majority in the House of Representatives was to cut taxes beyond what was called for in the earlier balanced-budget agreement. But then there are the spending cuts needed to achieve balance. The Congressional Budget Office did not produce the numbers needed for this to work out, which prompted the Republican leadership to address a letter to the appropriations subcommitee warning that if the CBO did not begin to produce better numbers, "we must review [its] structure and funding" [24].

FROM POLITICALLY INCORRECT TO NOBEL PRIZE

The seed from which the mathematics discipline called linear programming first germinated was planted in the late 1930s when the Leningrad

Plywood Trust approached the mathematics and mechanics department of Leningrad University for help in solving a production scheduling problem of the following nature. The Plywood Trust had different machines for peeling logs for the manufacture of plywood. Various kinds of logs were handled, and the productivity of each kind of machine (that is, the number of logs peeled per day) depended on the wood being worked on. The problem was to determine how much work time each kind of machine should be assigned to each kind of log in order to produce the largest number of peeled logs. A basic condition which had to be satisfied was that if logs of a given type of wood, oak, let us say, made up a specified percent of the input, 5 percent, for example, the peeled oak logs should also make up 5 percent of the output.

The germination of this seed is due to Leonid Kantorovich who saw that it, together with a wide variety of economic planning problems, could be formulated in terms of math problems called linear programs. These problems involved the optimum distribution of work time of machines, minimization of scrap in manufacturing processes, best utilization of raw materials, optimum distribution of arable land, optimal fulfillment of a construction plan with given construction materials, and the minimal cost plan for shipping freight from given sources to given destinations.

In 1939 Kantorovich published a report [25] on his discoveries which included a method for solving all the linear programs he had formulated for the aforenoted problems. The chaos of the Second World War and the postwar intellectual climate in the Soviet Union did not allow for the development and implementation of Kantorovich's linear programming methods in the Soviet economic scene. Any work which seemed to deviate from Marxist-Leninist ideology as interpreted by Joseph Stalin's closest ideologues could have chilling consequences for its authors. Kantorovich's mathematical methods for economic planning problems were neutral as far as Marxist-Leninist thinking was concerned, but they were new and a radical departure from comfortable orthodoxy, which make them suspect. This was enough to put them into a deep freeze.

The thaw in the Soviet Union's intellectual climate which followed Stalin's death in 1953 saw the rebirth, development, and implementation of Leonid Kantorovich's linear programming methods into the economic life of the USSR. Independently of the Soviet scene, linear programming methods were developed in the United States and Western Europe in the late 1940s, and the 1950s and 60s saw the development of a wide variety of linear programs for problems arising in such areas as economic planning, accounting, banking, finance, industrial engineering, and marketing.

In 1975 Kantorovich was a corecipient of the Nobel Prize in economics for his development of linear programming methods and their application to economic planning.

CONTRADICT ESTABLISHED ORTHODOXY AT YOUR PERIL

President Bill Clinton's proposal to raise the minimum wage from $4.25 to $5.15 an hour in 1996 was denounced by House Majority Whip Tom DeLay as "a job killer cloaked in kindness." It is, after all, a variant of the economic principle of supply and demand. Raise workers' wages and you increase the price of labor, companies will respond by hiring fewer workers.

But is it that simple and straightforward a proposition: no ifs, ands, or buts? Princeton economists David Card and Alan Krueger undertook a study of a situation which arose in 1992 when New Jersey raised its minimum wage but neighboring Pennsylvania did not. Orthodox economic theory would have it that New Jersey would lose jobs relative to Pennsylvania. Card and Krueger studied fast food restaurants in both states, using those chosen in Pennsylvania as a "control" group, and found that there was no statistical difference in fast food employment in the two states [26]. Card and Krueger's numbers were staggering in their implication: moderate minimum-wage increases have no impact on jobs.

A spirit of honest, open-minded inquiry would prompt further analysis and studies of what factors would account for this finding and what strings would have to be attached to a fundamental supply-and-demand principle of economics to make it better fit reality's behavior. This, however, was far from the immediate reaction. The Clinton administration was pointing to results like those obtained by Card and Krueger to support raising the minimum wage. For the camp strongly opposed to this, and for those who saw a principle of classical economics they considered inviolate under strong attack, the answer was polemics.

The Card-Krueger study and its authors, viewed as the lead dominos, were subjected to the heaviest assaults. If they fell, then the rest would immediately collapse, or so this version of the domino theory held. The counterattack took the form of articles in such influential publications as *The Wall Street Journal*, *The Washington Times*, and *Business Week* which took issue with the methodology of the Card-Krueger study and the reliability of the data presented. Ordinarily, such might be viewed as fair academic exchange, but the tone and thrust of the attacks make clear that it was far from that. One writer went so far as to take issue with the American Economic Association's award of the John Bates Clark Medal (given every two years to the most outstanding economist under 40) to David Card in 1995, suggesting that the quality of the review process was faulty.

As Alan Blinder, professor of economics at Princeton and formerly vice chairman of the Federal Reserve, put it: "Professors David Card and Alan Krueger are among the most careful empirical researchers in the economics profession" [27]. After all the huffing and puffing the lead dominos stood, and Card and Krueger emerged with reputations enhanced rather than diminished. [See also Ref. 28.]

5

Numbers, Numbers, Everywhere

ARE WE ON THE "RIGHT" NUMBER TRAIL?

Desiring a sense of how tall the freshmen players were, Angela Roberts, coach of Huxley College's women's basketball team, sent her assistant coach, Janice Reid, to find out. Janice weighed the players and presented the results to Angela, who was not satisfied.

"What's the problem," asked Janice, "I used the new scale we obtained and there's no question about the reliability of my numbers."

"I'm not questioning the reliability of your numbers," replied Angela, "but how well chosen your numbers are considering that my interest is in the heights of the players."

"You said that you wanted a sense of how tall the players were. Their weights give you such a sense because, in general, height is proportional to weight."

"Not good enough; your numbers give me too crude a sense of how tall the girls are to be useful. Why don't you just measure their heights directly?"

Having just learned from Metro Weather that the temperature was 60°, Richard Browne put on his jacket and stepped outside, intending to take a walk. Two minutes later he was back inside, shivering. He turned on the Weather Channel and learned that the wind chill reading was 28°. That explains it, he thought, I was looking at the wrong number.

The Baldwin Insurance Company hired marketing analyst Bob Williamson to develop a strategy to encourage safe driving and to make its insurance policies more attractive. Bob suggested that Baldwin offer a 5 percent discount to policyholders who had been with the company for five years and had been accident free. An additional 1 percent discount would be given for each additional year that had been accident free up to a maximum of 12 percent.

"The virtues of number of years accident free are that it's simple to determine and it's accurate," Bob pointed out to his associate Joe Pesky.

"Its downside is that it's not reflective of safe driving; it doesn't take into account the number of miles driven," Joe observed. "My cousin has had a driver's license for over 30 years and he's never been involved in an accident. If Baldwin were giving out awards for the largest number of years accident free, he'd be a prime contender. The only thing is, I don't think he's driven a hundred miles. He hates to drive and retains his license for identification."

"Don't cancel the course in modern Lithuanian literature," said Albina Jurkis, chairwoman of the East European languages department of Ecap College, to Andrei Markov, dean of academic affairs at the college. "The course registration increased by 100 percent over the last week and with such rapid growth I'm optimistic that we'll have a strong registration."

"One hundred percent growth sounds very impressive," replied the dean, "but what you're talking about is a course registration that has gone from one to two students. This can hardly be characterized as rapid growth."

The long-time standard measure of an airline's reliability is its percentage of on-time arrivals, where a flight is deemed on time if it arrives within 15 minutes of its scheduled arrival time. Such data are widely trumpeted by airlines in their advertising campaigns. But is this figure a suitable measure of an airline's reliability? According to Julius Maldutis, an airline analyst with Salomon Brothers, the answer is no. Maldutis argues that a much better measure of reliability is the percentage of flight miles completed. Look at the cancellation rate, which is indicative of a more troublesome situation to travelers than that indicated by the artificial on-time statistic, says Maldutis [1].

Unemployment has been regarded as the classic indicator of inflation. But a study by Stephen Cecchetti, Rita Chu, and Charles Steindel while at the Federal Reserve Bank of New York concluded that some 85 percent of the time the unemployment rate produced worse forecasts about inflation than those assuming that inflation would continue on its current path. This analysis is based on data from 1985 to 1998. What about

MI (currency plus checking and savings accounts)? MI was off the mark some 92 percent of the time [2,3].

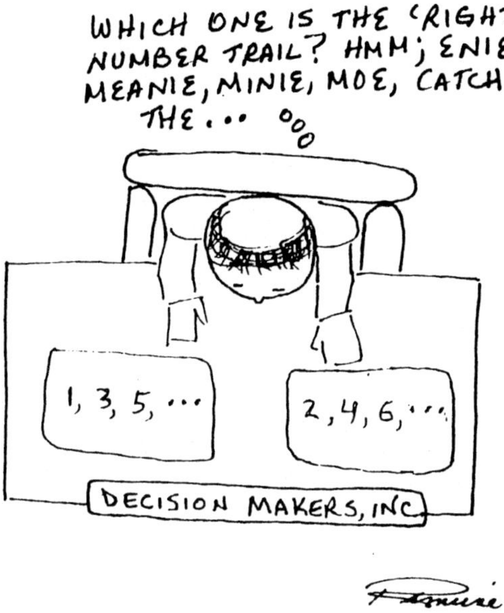

A study to be undertaken prompts the call to get the numbers. But which numbers would be appropriate? The answer, of course, depends on circumstances. Sometimes the answer seems "obvious" (as in the case of Angela Roberts desiring a sense of how tall her freshmen players were), but "obvious" is a judgment call. What's obvious to one may not be obvious to another, nor to the same person at another time.

A VISIT TO HUXLEY COLLEGE

Dr. Ivor M. Wisdom, a financial analyst, was hired by President Julius Marx of Huxley College to analyze the operations of the departments of Huxley College and make recommendations on how to improve their financial efficiency. Dr. Wisdom defined the income of each department as the tuition income of the students being serviced by the department minus salary cost. He collected data on the class size of each instructor and each instructor's rank and salary and found that a number of full professors at the top of the salary scale were teaching classes with a small number of students. To improve the income of the departments, Dr. Wisdom recommended that teachers at the top of the salary scale be assigned basic level courses, which can be expected to have a large number of students.

There is no question that Dr. Wisdom's data are consistent with his view of financial efficiency and reliable, but are they well chosen considering the issue? "Horsefeathers," cried Dean Arthur Marx, President Marx's brother. "It's irrelevant to the overall financial efficiency of the departments and Huxley as a whole. You could change the teaching assignments at the last minute and this game of academic musical chairs will not change a department's overall tuition revenue or cost, and these factors determine financial efficiency. The scheme is counterproductive in that it deflects us from the real issue of financial efficiency. It may also have negative academic consequences by leading us to assign courses to be taught on the basis of an instructor's spurious personal efficiency rather than academic qualifications. Go back to the drawing board Wisdom and come up with something better."

Much more is involved than what might at first appear to be an esoteric argument taking place in the clouds hovering about Huxley. The institution's success in improving its financial efficiency is at stake. The quality of Huxley's education in terms of how faculty teaching assignments are determined is also at stake. The outcome of this argument will be felt most by Huxley's students.

MEASURING THE NATION'S WELL-BEING

What would serve as a suitable measure of how well Americans are doing? Many would say the gross domestic product (GDP), which is a measure of the value of all goods and services. Some would disagree, arguing that factors having a negative impact on our well-being are not taken into account by GDP.

Table 5.1 GDP and GPI Values (billions of 1992 dollars)

	Gross domestic product	Genuine progress indicator
1970	3398	1781
1980	4615	1982
1990	6136	1965
1998	7552	1770

Source: Ref. 4.

Redefining Progress, a San Francisco think tank, has proposed a "genuine progress indicator (GPI)" as a measure of how well we're doing. GPI deducts the costs of such things as divorce, crime control, pollution cleanups, and income inequality [4]. Table 5.1 shows GDP and GPI values in billions of dollars in terms of 1992 dollars.

A PRECURSOR OF ECONOMIC COLLAPSE?

Several years ago the gross national debt of the United States, it was generally agreed, was in the neighborhood of $4.2 trillion, a figure which boggles the mind. If you had to transfer this amount of money in $1 bills from one location to another, you would have to deal with a stack of bills 267,000 miles long, which is roughly the distance from Earth to its moon plus the size of Earth's circumference.

The figure sounds ominous, no doubt, and one point of view is that it and the rate at which it had been increasing portend catastrophic consequences in the offing. When the debt grows faster than the country's ability to carry it, a breakdown with social, political, and economic upheaval is inevitable, and we are coming dangerously close to this state, this view has it. Its proponents include former Senators Paul Tsongas and Warren Rudman, who formed the Concord Coalition to rally public support for making debt reduction a top priority, James Davidson, founder of the National Taxpayer's Union, and Ross Perot, who made debt reduction the major theme of his 1992 presidential campaign.

Another view argues that in terms of the state of economy, we are looking at the wrong figure and that, while debt reduction is desirable, it should not be given top priority and carried out in a "mindless" way since this will severely damage the economy. Its proponents focus on the ratio of publicly held debt to gross domestic product, at that point 51 percent, which they argue is not a figure to be alarmed over. The major problem is the low productivity growth rate, this view argues, and that budget cutting will make impossible to do what had to be done in the way of education and infrastructure repair to enhance productivity. This camp includes Lester Thurow, former dean of the Sloan School of Management of the Massachusetts Institute of Technology, and Robert Eisner, professor of economics at Northwestern University and past president of the American Economic Association. Among many writings on the subject, see Refs. 5 to 7 for a more complete discussion of these points of view.

The issue was reignited when President Bill Clinton proclaimed, "Let's make America debt free for the first time since 1835," in his January 2000 state of the union message. Eliminating the debt shouldn't be the nation's goal, critics responded. The ratio of public debt to gross domestic product, which has been falling for several years, is the number to watch, they countered. Vice President Gore, seeking his party's nomination for president, jumped on the bandwagon by declaring that we should continue to pay down the debt even when the economy slows. "He should wash his mouth out with soap," responded Nobel laureate economist Robert Solow. This is much more than an esoteric argument over an economic point. Big money is at stake and how it is spent will affect the lives of millions of people. If the first priority is to pay off the debt, other needs—education, health, infrastructure, tax relief, you name it—must take lower priority. (For discussion of what's behind the Gore camp's thinking on the debt, see Refs. 8 and 9.)

The history of debt reduction in the United States is worthy of note. While the observation that the United States was virtually debt free in 1835–1836 is correct, it is also true, as Prof. Frederick C. Thayer points out, that a serious depression began in 1837. "Every sustained period of debt reduction in United States history has been immediately followed by a depression. There have been six such periods, preceding the economic collapses of 1819, 1837, 1857, 1873, 1893 and 1929. For whatever reasons, systematic debt reduction has never led to prosperity" [10].

WHAT'S THE "BEST" MEASURE OF THE IRON HORSE?

Injuries could not stop him and illness did not deter him. Yankee great Lou Gehrig played on and on in 2130 consecutive games, earning him the title "The Iron Horse." Gehrig's record for durability and endurance stood for over half a century until it was broken by Cal Ripkin, Jr. on September 6, 1995.

But is the number of consecutive games played the "best" measure of durability and endurance? The first question is: Does a player have to play the entire game to receive credit for having played in the game? No, only an appearance in the game is required, which prompts the question: Would the number of consecutive games played without missing an inning be more reflective of durability and endurance? From this point of view Ripkin beat Gehrig's record long before September 6, 1995; he played over 900 games without missing an inning, whereas Gehrig played consecutive games without missing an inning in only one season.

What about position played? Some positions inflict more wear and tear on a player than others. Should this be taken into account by a measure of durability and endurance? Many would say yes, but what an appropriate way of doing this would be is not at all clear. On this count Ripkin, who played the punishing position of shortstop, has it over Gehrig, who played the less grueling position of first base.

But then, should the number of plays per position be taken into account? Gehrig at first had more plays than Ripkin at shortstop because the entire infield threw to Gehrig.

Is the number of consecutive games played the "best" measure of a player's durability and endurance? From the preceding considerations, many would say no. This measure could be refined and sharpened to better reflect durability and endurance, but does it make sense to do so? The answer depends on what one seeks to accomplish. This number does reflect a player's durability and endurance, and it's straightforward and simple (no troublesome qualifications and conditions). This is a hard combination of positive features to convincingly improve on.

The title of this section, What's the "Best" Measure of The Iron Horse?, might suggest that there is a "best" measure, one which is perfect and cannot be improved on. The preceding discussion suggests that the idea of a "perfect" numerical measure of a player's durability and endurance is illusionary. Many such measures can be constructed, depending on which features of durability and endurance are to be taken into consideration, and one can argue the merits of one against another. This is the best that we can realistically hope for.

6

There's Less to It Than Meets the Eye

Numbers are neutral in the sense that they themselves do not carry a built-in one-and-only-one interpretation. How numbers are interpreted will, at best, depend on the capability and judgment of the interpreter and, at worst, on the cleverness of the spin doctor entrusted with putting on them the best possible face. It is often the case that "experts" see very different things in the same numbers and that different numbers concerning the same issue seem to be contradictory.

STATISTICAL LINK VS. CAUSE

A study of men under 55 published by *The Journal of the American Medical Association* in February 1993 concluded that baldness on the top of the head, called vertex baldness, is correlated with a slight but definite increase of heart attack. The more extensive the vertex baldness, the higher the risk, the study found. Does this mean that vertex baldness causes heart attacks? *No*! What had been found is a statistical link between two factors. A statistical link between two factors does not imply cause and effect. It's possible that some third factor influenced either or both variables or that the correlation is spurious.

There is a statistical link between the number of beds in a city and the city's death rate. This does not mean that lying in bed is a cause of-death. A bed is just where many of us find ourselves when we pass on. There is a statistical correlation between Scholastic Aptitude Test score and academic success to the extent that SAT score is used by many colleges as part of their admissions profile. The SAT score is not a cause of academic success. At best there are behind-the-scene characteristics that influence the SAT score and academic success in such a way that one might be used as a "successful" prediction tool for the other.

Before making a major change in life style, what can we do to weigh the evidence in an age when alarming reports of health risks connected with almost every aspect of life have become commonplace? The following guidelines are useful, but only as guidelines.

1. Keep in mind that statistical correlation by itself does not imply cause.
2. Consider the source of the report and the publisher. There are no guarantees, but it is reasonable to treat a study done at a major university, such as Harvard or Yale, as more authoritative than one coming from an unknown school. Reports published by peer-reviewed journals, such as *The New England Journal of Medicine*, are more reliable than those that come out of medical meetings or news conferences. An editorial should accompany major studies discussing their strengths and weaknesses.
3. Are there other studies which have reached the same conclusion?
4. Is the finding supported by animal studies?
5. Are there biological explanations for the observed association?

DOES CIGARETTE SMOKING "CAUSE" CANCER?

When teenager Michelle Lewis took up smoking, her mother became concerned about what this would do to her health. "Mom, I know there is a strong statistical correlation between cigarette smoking and lung cancer, but this does not establish that cigarette smoking causes lung cancer." You're right, Michelle, this statistical correlation does not by itself establish, beyond a reasonable doubt in a scientific sense, a causal relationship.

The progression from strong statistical correlation to beyond a reasonable doubt takes place in a series of steps, some big and some small. In 1952 Dr. Richard Doll established that the risk of developing lung cancer is directly proportional to the number of cigarettes smoked. In 1953 a group headed by Dr. Ernst Wynder, working with mice, provided experimental proof that the contents of cigarette smoke could induce cancers in skin; 44 percent of the mice that had the exposed skin on their backs painted with tobacco tars taken from cigarette smoke developed malignant tumors. A pivotal link between smoking and human cancer was established by the pathologist Dr. Oscar Auerbach and several colleagues who, over an eight-year period, examined over 100,000 slides of lung tissue taken from over 1500 deceased men and women. They found high incidences of cancerous or precancerous cells among the tissues of those who had smoked, with the degree of cell damage proportional to the amount of smoking as determined from interviews of relatives. Dr. Auerbach's findings played a basic role in the 1964 Surgeon General's report on the danger of smoking.

As to what happens at the cellular level, until recently the technology needed to address this question wasn't available, but in 1996 research teams led by Dr. Moonshong Tang at the M. D. Anderson Cancer Center in Houston and Dr. Gerd Pfeiffer at the Beckman Research Institute of the City of Hope in Duarte, California, established that benzo (α) pyrene, a chemical in cigarette smoke, causes damage to a protein in lung cells known as p53 that is identical to that found in many patients with lung cancer. The p53 protein acts to suppress the runaway growth of cells that lead to tumors, and the damage done to p53 renders it incapable of carrying out this task.

With this link an airtight case that cigarette smoke is a cause of lung cancer is made, no ifs, ands, or buts about it. What remains to be better understood is why individual smokers differ in their tendency to develop lung cancer.

"You're a smart girl, Michelle," said her mother, "and I hope that you're smart enough to trade in the idea of being 'cool' for the benefit of a healthy life before you're hooked."

IT'S DANGEROUS TO BE A DEMOCRAT?

Burt Jones is giving serious thought to joining the Democratic party. His friend Mike thinks it's a bad idea. "Burt, consider the following data reported by Richard Morin based on survey data obtained by the University of Chicago's National Opinion Research Center between 1972 and 1993" [1].

1. *Democrats are poor*. The average income of Democrats was 25 percent below that of Republicans.
2. *Democrats have unpleasant habits*. Thirty-seven percent of them smoke, as compared to 30 percent of Republicans; 23 percent of them have seen an X-rated movie within the past year, as compared to 17 percent of Republicans.

3. *Democrats are unhappy.* Thirty-one percent of them describe themselves as very happy, as compared to 39 percent of Republicans."

Burt's first reaction was to question the legitimacy of these figures, but then he realized that there is a larger issue that should be focused on. Accompanying them is the suggested conclusion that if you're a Democrat, you're more likely to acquire these characteristics than if you're a Republican. A correlation is being suggested as cause and effect. It is also being assumed that since more Democrats exhibit the aforenoted characteristics, they do so because they are Democrats.

Burt decided that the decision on whether to join the Democratic party or not should be made irrespective of this data.

DO RANKINGS SAY IT ALL?

The eldest child in the Allison family will be ready to start college next year, and more and more the Allisons' thoughts are turning to the question: What is the "best" school? Like many concerned parents and prospective college students, this may prompt the Allisons to reach for the most recent *U.S. News & World Report's* annual "America's Best Colleges" issue which states a numerical pecking order for colleges and universities.

One of the most useful features of numerical values is that they can be employed to communicate a sharp, unequivocal distinction which is to be made. If a horse race is run and Jupiter wins followed by Ally followed by Candy, we would assign 1 to Jupiter, 2 to Ally, and 3 to Candy. There is no danger of misunderstanding what 1, 2, and 3 mean in this context.

But is it meaningful to rank colleges and universities in such a manner? Rankings of this sort suggest a level of precision that is illusionary for colleges and universities, many would argue. In a letter to the editor of *U.S. News & World Report*, Gerhard Casper, president of Stanford University, wrote [2]:

> As the president of a university that is among the top-ranked universities, I hope I have the standing to persuade you that much about these rankings—particularly their specious formulas and spurious precision—is utterly misleading. . . . I am extremely skeptical that the quality

of a university—any more than the quality of a magazine—can be measured statistically.

As an example, Casper cites "Value added":

One place where a change was made openly was, perhaps, the most openly absurd. This is the new category "Value added." I quote the magazine: Researchers have long sought ways to measure the educational value added by individual colleges. We believe that we have created such an indicator. Developed in consultation with academic experts, it focuses on the difference between a school's predicted graduation rate—based upon the median or average SAT or ACT scores of its students and its educational expenditures per student—and its actual graduation rate.

This passage is correct that such a measure has long been sought. However, like the Holy Grail, no one has found it, certainly not the "we" of this passage. The method employed here is, indeed, the apotheosis of the errors of the creators of these ratings: valid questions are answered with invalid formulas and numbers.

Let me examine an example in "Value added": The California Institute of Technology offers a rigorous and demanding curriculum that undeniably adds great value to its students. Yet, Caltech is crucified for having a "predicted" graduation rate of 99 percent and an actual graduation rate of 85 percent. Did it ever occur to the people who created this "measure" that many students do not graduate from Caltech precisely because they find Caltech too rigorous and demanding—that is, adding too much value—for them? Caltech could easily meet the "predicted" graduation rate of 99 percent by offering a cream-puff curriculum and automatic A's. Would that be adding value?

. . . Were *U.S. News*, under your leadership, to walk away from these misleading rankings, it could be a powerful display of common sense. I fear, however, that these rankings and their byproducts have become too attention-catching for that to happen.

Could there not, though, at least be a move toward greater honesty with, and service to, your readers by moving away from the false precision? Could you not do away with rank ordering and overall scores, thus admitting that the method is not nearly that precise . . . ? Could you not, instead of tinkering to "perfect" the weighings and formulas, question the basic premise? Could you not admit that quality may not be truly quantifiable, and that some of the data you use are not even truly available (e.g., many high schools do not report whether their graduates are in the top 10 percent of their class)?

> Parents are confused and looking for guidance on the best choice for their particular child and the best investment of their hard-earned money. Your demonstrated record gives me hope that you can begin to lead the way away from football-ranking mentality and toward helping to inform, rather than mislead, your readers.

Casper's hope that we would be led away from football-ranking mentality has not been realized. This mentality has been expanded to encompass such enterprises as graduate and law schools, hospitals, and health maintenance organizations.

For further discussion of school rankings, see Refs. 3–12.

The Twentieth Century's 100 Best Novels?

With leisure time on his hands and a desire to catch up on his reading, but only the "best," Arthur turned to the Modern Library's 1998 ranking of the 100 best twentieth-century novels written in English for suggestions. If anyone knows, the Modern Library's 10 member board knows, thought Arthur. He found the list precise and for that reason reassuring. There's first place, second place, and so on down the line—no ifs, ands, or buts.

Spotting Arthur reflecting on the list, his friend David greeted him with some disturbing news. "The board members say that they never ranked the books; they were asked to check off books from a master list of 440 titles given them by the publisher, without putting them in any special order, noted David. Also, the judges did not agree on the features they were to consider and priorities—best-written books, the most important ones, or those that have been most influential. One judge was willing to confess that he voted for books he had not read; he believed they belonged somewhere on the list because of reputation or influence. Another board member, former Librarian of Congress Daniel Boorstin, expressed the view, 'Literature does not live in rankings'" [13].

"Are you implying that the precision of this ranking is illusionary?" asked Arthur. "My friend, you have broken the code," replied David. "What about other rankings," asked Arthur; "should the implied precision of all be considered suspect?" "That would be my recommendation," replied David. "Do not accept a numerical ranking at face value. If it's a matter of serious interest to you, check behind the scene to learn what the basis for the ranking is and keep in mind that for some things rankings are meaningless."

LORK FROM MORK

Concerned about charges of subtle patterns of bias at its executive levels, the United Federation of Worlds set up a commission to investigate. During its hearings Lork from Mork pointed out that Morkians are 30 percent of the Federation's workforce at its lower levels, but that they make up only 1 percent of its executive staff. "Good faith recruitment efforts have been made," observed Lork, "but the statistics show subtle patterns of discrimination against Morkians."Tallia from Talos I commented: "Look at the Universe Games held every hundred years. For the last thousand years 75 percent of the participants chosen have been Morkians. Does this mean that there has been a pattern of bias that favored Morkians over the last thousand years?" "Certainly not," answered Lork, "They earned the right to be there."

Do the statistics establish patterns of discrimination against and for Morkians in these cases? There is a fundamental distinction to be observed between numbers, which by themselves are neutral and have no bias, and the interpretation given them, which may vary considerably from person to person.

The percentages cited by Lork and Tallia make clear that there are discrepancies. Discrepancies are analogous to potential symptoms of disease, but their appearance may or may not indicate that discrimination has taken place; further investigation is required. It should also be kept in mind that the absence of numerical discrepancies does not, by itself, necessarily mean that discrimination has not taken place.

MATH TO THE RESCUE?

The following message from the dean of administrative affairs of Ecap College of Mork University was sent to all department chairpersons and faculty of the College.

> Dear Colleagues:
>
> I have placed a high priority on resolving internal equity issues related to faculty salaries in Ecap College. To this end, I have been working with the Human and Non-Human Resources Department for the last several months.
>
> A regression analysis that employed many variables (e.g., rank, discipline, performance, gender, and years of service) was conducted by Human and Non-Human Resources. The analysis was helpful to identifying certain Ecap faculty members who qualify for salary adjustments.
>
> With the support of the President, the Vice President for Academic Affairs, and the Vice President for Human and Non-Human Resources,

we were able to provide midyear salary adjustments to those who qualified. These colleagues have already been notified. The overall effect was to raise the salaries of one quarter of the College's full-time faculty and to increase the College's full-time salary base by $100,000.

Of course, I will continue to address the salary and other needs of the College. With support and cooperation at all levels, there is every reason to be optimistic about our future.

Please accept my sincere appreciation for your many contributions to the College. They are moving us to a higher level of quality and recognition, for which I am grateful.

The dean of administrative affairs employed a math tool (called regression analysis) to identify inequities, which in turn prompted the salary adjustments made. The question that arises is: Can a math tool (regression analysis or whatever), by itself, identify inequities? The answer, of course, is no. This amounts to the same basic scenario played out by Lork and Tallia, with an air of higher sophistication being communicated by the term regression analysis rather than the more mundane percentage. Regression analysis will identify differences, but whether these differences translate to inequities is a judgment call which the dean preferred not to make.

The dean's message prompted the following reply from the chairman of the mathematics department.

To: Dean of Administrative Affairs, Ecap College
From: W.J.A., Chairman, Mathematics Department
Re: Don't Make It Complicated When You Can Make It Simplistic

Employing regression analysis to "resolve internal equity issues relating to faculty salaries in Ecap College" is a stroke of genius. When I think of the countless hours that I spent as a member of the Salary Review Board and as Chairman of the Mathematics Department struggling with the nuances distinguishing salary differences from inequities, I could kick myself. If I had been smart enough to think of regression analysis 20-odd years ago, I could have saved enough time to write another book.

Just to illustrate how foolish I had been, I confess to having given serious thought to matters such as the following:

Professors X and Y have ten years of teaching experience, let us say. Are they comparable or does X have one year of teaching experience repeated ten times while Y shows ten years of growth?

Professors X and Y have ten years of teaching experience. Are they comparable when I seek to assign X to a course where he would be most effective, while I seek to place Y where he would do the least damage?

Professors X and Y have ten years of teaching experience. How "insightful" is the one in explaining course material as opposed to the other? Can the students tell the difference?

How does the spectrum of courses I can comfortably assign to X compare with what I can comfortably assign to Y?

X has published one paper in the last four years, Toward a Solution of the Riemann Hypothesis Conjecture, which has been praised as a major development in the field. Y has published four Mickey Mouse papers in the last four years. How do these publication records compare? Now I can appreciate the simplicity of it all; it's 4 to 1.

An advantage of regression analysis for you is that there's no need to solicit chairpersons' judgment on matters of salary difference and equity. Once regression analysis speaks, that's it. Chairpersons can properly devote themselves to attending meetings, and trying to sound cooperative, cordial, and awake.

Another beautiful dimension of a regression analysis approach is that it doesn't leave fingerprints, at least obvious ones. A mathematical analysis presents an illusion of objectivity and precision which most people find intimidating. This helps you to keep them at bay and allows you to avoid taking direct responsibility for unpopular decisions.

7

Lessons from Chapters 1 to 6

MADNESS OVER MATH

It wasn't too long ago that if you gave a talk during which you spoke utter nonsense, the chances were good that you would be denounced and hooted off the stage. On the other hand, if you had the talk committed to print and published, it took on a new life, being viewed as sacred and quoted and cited in books that claimed to be purveyors of wisdom.

These days the number has replaced the printed word as a sacred object and the more startling the number, the greater the wisdom that must lie behind it is a far too widely accepted view. While this might sometimes be true, the cases considered in the preceding chapters should serve to make clear the variety of ways that cited numbers may fail to live up to expectations.

What then can we do to protect ourselves from being "taken" by numbers? I suggest developing a proper perspective about them and asking the right questions. Consider:

NUMBERS IN PERSPECTIVE

Do the following views express math insights, math myths, or a bit of both attractively packaged as mathematical Trojan horses?

1. *A presentation supported by figures is more credible than one that is not.*

This is an attractively packaged mathematical Trojan horse. The basic qualification that the figures be reliable and well chosen must always be kept in mind.

2. *Numbers are neutral. They are not affected by cultural, economic, and political differences that influence people and therein lies their strength. 1 + 1 = 2, for example, is universal for all peoples, countries, economic and political systems.*

Another attractively packaged mathematical Trojan horse. As a statement of mathematics without regard to background or context, there is no question that 1 + 1 = 2, and that numbers are neutral in this sense. The complication occurs when we inquire about how the first 1 was obtained, where the second 1 came from, and how the resulting 2 is to be interpreted. When questions of what context gave rise to the numbers, and how the result of the mathematical operation (addition, in this case) is to be interpreted, we are taken back to questions of reliability of the numbers and how well chosen they are to the study being undertaken.

If we purchase a container of milk for $1 and another container of milk for the same price, then the total cost to us is $2, no question about it. If we are informed that a test of a weapons system conducted on March 1 was successful, as well as the one conducted on March 5, it follows that the two tests were successful. This 2 is not as well anchored in reality as the aforenoted one because its realism is dependent on the reliability of the data we started with. (See "Indefensible Defense Figures," Chapter 1.)

3. *Numbers give weight to a view through the sense of precision they communicate, thereby advancing it to a plane which commands recognition, respect, and acceptance.*

Another attractively packaged mathematical Trojan horse. The hook in the bait is with the word acceptance. Acceptance is conditional on the numbers being reliable and well chosen, and so is respect.

HOW WELL CHOSEN ARE THE NUMBERS? HOW RELIABLE ARE THEY? WHAT INTERPRETATION IS "APPROPRIATE"? TEST QUESTIONS

Application of the following 11 questions to an article or presentation which relies on numbers should put us in a better position to come to grips with misleading or fraudulent figures and questionable interpretations.

Test Questions

1. *Are the numbers well chosen considering their intended purpose?*

This is a key question that should always be kept in mind. It is a difficult one to address in general, but if we are prompted to answer in the negative, questions concerning their reliability become moot.

2. *Is a source given for the number(s)? If not, try to track it down.*

If the answer is no, then questions 2, 3, 4, and 8 are not viable. Be on guard. If you succeed in locating a source, then apply 2, 3, 4, and 8. In the final analysis, the burden of "proof" of the number's credibility is in the author's hands, not yours.

3. *Is the source unbiased and reliable?*

What experience have I had with the source in the past? Has it earned membership in the Slippery Statistics Society, a reputation for reliability, or is its reliability status unknown?

4. *How were the numbers obtained?*

Does the method for obtaining the numbers have weaknesses which may compromise their integrity?

5. *How current are the numbers?*

If they are stale, they may no longer be applicable.

* D. Leonhardt, "A Tale of 2 Totals in the Jobs Report," *The New York Times*, Jan. 28, 2001, Bu-4.

6. *Is there more to the issue than the numbers are being interpreted as meaning?*

There often is; the question is, how is this dimension to be reconciled with the need for a realistic overall perspective?

7. *Are there other numbers which support the conclusion reached?*

8. *Are there other numbers which contradict the conclusion reached? If so, how are the contradictions to be resolved?*

9. *What assumptions underlie the numbers and conclusion obtained from them? Are these assumptions realistic?*

If the assumptions are unrealistic, the conclusions obtained from them must be considered suspect in terms of hitting reality's mark.

10. *Do the numbers admit contradictory interpretations? If so, how are these contradictions to be resolved?*

Also of interest is the question: What follow-up courses of action are suggested by the interpretations?

11. *When monetary amounts are being compared, is the comparison being made with respect to similar circumstances (time frame, for example)?*

If not, the comparison may be deceiving. Fifty million dollars spent in 1990, for example, cannot be equated to $50 million spent in 2002. Inflation must be taken into account for a realistic comparison.

POLL CONSIDERATIONS

Polls have become so numerous that Russell Baker's suggestion that "what this country really needs is a poll-ban treaty" [1] begins to sound more and more appealing, especially as election time approaches. Even if polls strongly supported this suggestion, they are likely to become more rather than less numerous. The problem centers on what we can do to help us distinguish polls that have been properly carried out from their flawed cousins.

The first three of the following points are applicable to polls in general.

1. Context and Basic Information. Focusing on poll results without an appropriate context is misleading. News accounts of poll results should give information about the date the poll was taken, sample size, survey design, percentage of respondents among those contacted, response options, random sampling error and what it means. Out-of-context poll results are best viewed with questioning skepticism.

a. Date the poll was taken. Are the poll data still timely or should they be considered stale?

b. Sample size. It's difficult for an onlooker to make a judgment about whether the size of the sample chosen is appropriate. Generally, other things being in order, larger sample sizes yield more precise estimates, but a large sample size cannot compensate for a sample that is not properly chosen (which has to do with survey design) and a low response rate [see (d)].

c. Survey design. This has to do with how the sample is selected, which in turn depends on the purpose of the survey. If the purpose is to project a poll's findings onto a larger group, and a measure of the sampling error is desired [see (f)], then the sample should be chosen at random (that is, in an unbiased manner) from that group.

d. The Response Rate: Polling's Unseen Achilles Heel. A major lesson of *The Literary Digest*'s 1936 presidential poll is that a low response rate renders the results obtained untrustworthy as a basis for predicting the attitudes of the target population. (See "Polling and Its Perils," Chapter 3.)

The explosive growth in poll taking and telemarketing, which competes with polling for the public's time and patience, has made it increasingly difficult to achieve adequate response rates. With careful preparation of the groundwork (which might include sending advance letters to the sample of people chosen to be polled announcing that a poll taker will be calling and explaining the poll's purpose) response rates of 80 percent or better have been achieved, but this requires time and proper budget, which tend to be in short supply. The reliability of overnight polls, whose response rates are often in the teens, is particularly open to question.

Perhaps not surprisingly, polling firms generally do not release their polls' response rates.

e. Response options. The larger the number of response options (within reason), the more accurate the assessment of views solicited can be expected to be.

f. Random sampling error. There is a difference between the views held by members of a sample and those held by the population the sample is chosen from. The random sampling error is a measure of the extent to which the results obtained from a sample might differ from those obtained from the population itself. If a poll shows that candidate A is favored over B by 52 percent to 48 percent and the sampling error is cited as ±2 percent, then a poll of the same size taken at the same time might have yielded 52 ± 2 percent in favor of A and 48 ± 2 percent in favor of B, or anything in between.

2. Questions. The complete wording of questions should be provided. We may then ask ourselves: Are the questions clearly posed? Are there leading questions whose coloring would favor a certain kind of response? Are there very personal questions that a respondent might be reluctant to answer truthfully?

3. Self-selected Respondents. A popular, but seriously flawed, polling technique is the mail-survey, magazine, online, or telephone poll in which the public is invited to respond to a written questionnaire, or call one number to register approval of a candidate or position and another number for disapproval. Polls of this sort lend themselves to gross manipulation by individuals and pressure groups and give us no handle on the opinions of nonrespondents. Self-selected respondents represent only themselves.

Any claim or suggestion that results obtained from polls of this sort express the views of a wider population is totally without merit.

4. *On-line Polls.* The first stage of on-line polling does not differ from mail-in or telephone-response polls. Pseudo-poll would be a more appropriate label for such techniques rather than poll, which has come to suggest a rigorous, scientific framework not possessed by the first stage of on-line polls and their ilk.

The advantage of online pseudo-polls over their kind is that the internet dimension permits the development of this pseudo-poll into a "legitimate" poll, sometimes termed an interactive online poll. This development is in progress. (See "The Future of Polling?" Chapter 3.)

When presented with the results of an on-line poll there is no way of knowing whether they are from a pseudo-poll or interactive online poll unless the methodology is identified. This is often not done.

Beware.

5. Who Commissioned the Survey and Who's Doing it? Many honest surveys are commissioned by interested parties, but there are fair-minded interested parties and not-so-fair-minded interested parties who are more concerned with manipulating opinion than obtaining an objective assessment.

It is useful to know who wants to know and who's doing the survey to help provide some perspective for the results obtained. The following points are applicable to a sequence of polls taken over a period of time on an issue, such as a sequence of preelection polls.

6. Consistency. It is useful to check several polls against each other. Are the results consistent? If not, try to find out why.

7. Watch for Trends, but Be Careful of Margins. If a polling organization takes a number of polls in sequence, do the results point to a trend? But if A leads B by 51 to 49 percent in one poll (for a margin of 2 percentage points) and in the next poll by 52 to 48 percent (for a margin of 4 percent age points), a 1-point shift in preferences has had a potentially misleading doubling in margins, from 2 points to 4 points. Moreover, if the accuracy gauge of the polls is cited as being ± 3 points, for example, the difference in the poll result might be due to sampling error, rather than to a shift in public opinion.

8. Popularity Polls. Popularity poll results for political figures should be considered suspect. Relative newcomers to a political contest suffer from a lack of name recognition, whereas established figures may have a record to run on which carries a number of negatives. Initially, voters' minds are not on an election which may seem distant. These percentages stabilize as election time approaches, but are still subject to the occurrence of events which none of the candidates have control over.

9. Push Polls. Beware of Push polls in which a campaign hires operatives to make calls spreading negative, misleading, and sometimes false information about an opponent while masquerading as pollsters.

During the Republican presidential nomination campaign in South Carolina in February 2000, Donna Duren reported to Senator John McCain that her teenage son Chris had taken a call from a pollster who characterized the senator as a "cheat and a liar and a fraud" [2]. Not subtle, but subtlety is not a requirement of push polling.

CONSIDER, FOR EXAMPLE, . . .

To the Editor

In a letter to the editor of *The New York Times* [3], Chu-Yuan Cheng argues that "the real threat of espionage seems largely overblown. China has only about 20 intercontinental ballistic missiles. The United States has about 7000 nuclear warheads; China has only 400."

Two questions arise. TQ-2 (Test Question 2): Is a source given for the numbers? As is usually the case with letters to the editor, op-ed articles and the like, the answer is no. This rules out TQ-3, 4, 5, and 9, but considering the highly sensitive nature of the numbers I would have to consider them highly suspect.

Assuming that you accept Professor Cheng's argument based on the numbers cited, would you do so if the correct numbers were substantially different? (TQ-6)

By the Numbers?

Based on a survey carried out by the group *Media Watch*, Betty Friedan pointed out several years ago that only 6 percent of the people on page 1 of *The New York Times* during February 1990 were females.

TQ-1 and TQ-6 come to mind. Is the number well chosen concerning the issue in question? Is there more to the issue than the number is being interpreted as meaning?

It is a straightforward, tedious exercise to verify the accuracy of the 6 percent figure. Mathematically speaking, it shows that there is a percentage difference between the number of males and females quoted on page 1 of the *Times* for February 1990; this much is indisputable. But, so what? What is the point of focusing on the front page of the *Times* in the first place to obtain this statistic? There is no defensible rationale for such a choice and in this we see its fundamental weakness.

Numerical differences are part of every situation, but the existence of such, cited out of context with trumpets blaring, does not by itself give them special significance. The 6 percent figure may seem startling, which is why it was trotted before us, the idea being to suggest bias on the part of the *Times*. It is an example of what *U.S. News & World Report* columnist John Leo termed a "grenade-like pseudo-statistic, blowing away all subtlety and leaving a thick residue of implied bias" [4].

Lies, Damned Lies, and . . .

Claim: Women earn 74 cents for every dollar men earn. *Interpretation and Conclusion*: Women are suffering a wage gap primarily caused by discrimination, which requires government action. George Will took issue with the 74 cents figure and its follow-up interpretation by, in effect, applying TQ-1, 6, and 8. Is the number well-chosen? Is there another number which contradicts the conclusion reached? How are the contradictions to be resolved?

Will calls attention to *Women's Figures: An Illustrated Guide to the Economic Progress of Women in America* by Diana Fuchtgott-Roth and Christine Stolba. Furchtgott-Roth and Stolba argue, according to Will, ". . . that the 74 cents statistic is the product of faulty methodology that serves the political agenda of portraying women as victims needing yet more government intervention in the workplace. . . . Between 1960 and 1994 women's wages grew 10 times faster than men's, and today, among people 27 to 33, women who have never had a child earn about 98 cents for every dollar men earn. Children change the earnings equations. They are the main reason that meaningful earnings contrasts must compare men and women who have similar experience and life situations" [5].

What is an interested onlooker who has neither the time nor the inclination to dig deeper (such as myself) to make of this? If confronted only by the 74 cent figure, I have cause for reservations in terms of general principles: (1) It's not clear what the foundation is for the 74 cent figure. This foundation, consisting of assumptions and qualifications—the small print, if you will—is all important to this figure's viability. (2) This figure has been seized by advocacy groups seeking to promote a social and political agenda. This, by itself, does not make the 74 cent figure incorrect, but it signals my antennae that there's a lot more to the issue than the figure might suggest.

From the point of view of advocacy groups, the 74 cent figure is too good not to be true. My interest is in seeing realistic numbers which can serve as a basis for realistic policy formulation. (3) Will's article on the book by Furchtgott-Roth and Stolba is sufficient to confirm my reservations about the 74 cent figure and the interpretation being given it by advocacy groups.

What Do You Make of This?

"The proportion of black and Hispanic students in New York City public schools who read and do math at the level expected for their grade is far below that of white and Asian students, new test results show" [6].

The numbers tell us that there is a difference between the proportion of black and Hispanic students who read and do math at the expected level and white and Asian students who read and do math at the expected level. As with the preceding three cases, that's all the math, by itself, can tell us. Now it's a matter of interpretation and TQ-10: Do the numbers admit contradictory interpretations? Do the interpretations suggest different courses of action?

Schools Chancellor Ramon C. Cortines saw the difference as due to deteriorating schools in the black and Hispanic communities, lack of qualified teachers, and insufficient teaching materials and technology. He saw investment in teacher training, technology, and rehabilitation of deteriorating school buildings as the path to equalizing the aforenoted proportions.

High school teacher Andrea Black Jeffries comments; "Your article [see Ref. 6] blames lack of teachers and computers as well as poverty. These do not explain the problem. We cannot teach a student who is not there. We cannot work with a parent who never attends open-school night. While we must examine 'what happens in places of learning,' we must also examine what happens before a student arrives there" [7].

Cortines' and Jeffries' accounting for the differences are not mutually exclusive, but the flow of resources to correct a troubling difference will depend on which interpretation is given most weight.

Junk Statistics?

In August 1996, Ross Perot addressed his Reform party: "Before our nation was formed the total cost to run our government was 67 cents per person a year. Then after our nation was formed, it went up to a dollar a year. Then 120 years later, in 1910, it was up to $6.75 per person a year. Then we passed the income tax law amendment in 1913, and 16 years later it had gone from $6.75 to $29 per person a year. Now, strap in your seat belt and don't eject, because you can't—today it's $5700 a year per person" [8].

One can verify that the numbers are accurate (take the federal budget at the times noted, divide by the estimated population, and round off).

The suggestion that we are spending ourselves into the grave has an ominous ring to it, but then we are challenged by TQ-11, 1, and 6: When monetary amounts are being compared, is the comparison being made with respect to similar circumstances? Are the numbers well chosen concerning the issue in question? Is there more to the issue than the numbers are being interpreted as meaning? The nation's circumstances, levels of sophistication, and the time frames chosen are so different that the comparisons offered are meaningless. The question, pointed out economist Charles Schultze, is "so what?" "If he [Perot] wants to say we have a serious, long-term problem, I completely agree," commented Mr. Schultze.

Would You Go to Bat for Baseball Statistics?

In his book *The Last Yankee: The Turbulent Life of Billy Martin,* David Falkner [9] concludes that Martin was the best manager of his era, possibly of many eras. Falkner's judgment was strongly influenced by baseball statistics compiled by the Elias Sports Bureau and a formula which claims to show which managers' teams won more games than they were reasonably expected to win.

TQ-1 and 6 arise. Are the numbers well chosen concerning the issue in question? Is there more to the issue than the numbers are being interpreted as meaning? No to the first, yes to the second, argues George Will.

In his review of Falkner's book, Will [10] disputes Falkner's interpretation which, he argues, the rest of the book refutes. "In fact," Will notes, "*The Last Yankee* might usefully be made required reading for graduate students in the social sciences and all others who need to be immunized against the seduction of numbers. . . . There are limits—and Mr. Falkner's reporting shows that Elias passed them regarding Martin—to the ability to capture messy reality in tidy formulas."

How Successful Was the Patriot?

During the Gulf War, television viewers were moved by scenes of Patriot missiles streaking across the sky to intercept Iraqi Scud missiles that had been launched against Israel and Saudi Arabia. The Patriot's success seemed to epitomize the success of a high-tech, low-causality military campaign.

Apart from very successful military public relations, how successful was the Patriot as a military tool? Different statistics have been given and, as is almost always the case, it is important to look further than the statistics if a realistic picture is to emerge.

The Army originally stated that Patriots "intercepted" 45 of 47 incoming Scud missiles, and President George Bush revised that to 41 of 42. What does this mean? Brigadier General Robert Drolet of the Army's Missile Command testified that "a Patriot and a Scud passed in the sky."

There are other statistics of interest. Before Patriots were employed in Israel, 13 Scuds fell near Tel Aviv. There were no deaths, but 115 people were wounded and 2698 apartments were damaged. After Patriots were employed in this region, 11 Scud attacks left 1 dead, 168 injured and 7778 apartments damaged [11, 12]. This is explained by the fact that successful hits led to more deadly debris being sprayed over a larger area than otherwise would have been the case and that the Patriots tended to strike the bodies of Scuds, leaving their warheads armed and able to cause significant damage.

Consider TQ-10: Do the numbers admit contradictory interpretations? Clearly they do; if so, how are these interpretations to be resolved? Both judgments and the winds of the spin doctors prevail.

The Reagan Economic Boom: Blessing or Disaster?

Martin Anderson, former advisor to President Ronald Reagan and senior fellow at the Hoover Institution, employs statistics to support his view that the Reagan economic boom was the greatest ever [13]:

The two key measures that mark a depression or expansion are jobs and production. Let's look at the records that were set.

Creation of jobs. From November 1982, when President Ronald Reagan's new economic program was beginning to take effect, to November 1989, 18.7 million new jobs were created. It was a world record: . . . The new jobs covered the entire spectrum of work, and more than half of them paid more than $20,000 a year. As total employment grew to 119.5 million, the rate of unemployment fell to slightly over 5 percent, the lowest level in 15 years.

Creation of wealth. The amount of wealth produced during this seven year period was stupendous—some $30 trillion worth of goods and services. Again, it was a world record. . . . According to a recent study, net asset values—including stocks, bonds and real estate—went up by more than $5 trillion between 1982 and 1989, an increase of roughly 50 percent. . . .

Income tax rates, interest rates and inflation. Under President Reagan, top personal income tax rates were lowered dramatically from 70 percent to 28 percent. This policy change was the prime force behind the record breaking economic expansion. . . .

The stock market. Perhaps the key indicator of an economy's booms and busts is the stock market, the bottom line economic report card . . . starting in late 1982, just as Reaganomics began to work, the stock market took off like a giant skyrocket. Since then, the Standard & Poor's index has soared, reaching a record high of 360, almost triple what it was in 1982.

There were other consequences of the expansion. Annual Federal spending on public housing and welfare, and on Social Security, Medicare and health all increased by billions of dollars. The poverty rate has fallen steadily since 1983.

When you add up the record of the Reagan years, and the first year of President Bush . . . the conclusion is clear, inescapable and stunning. We have just witnessed America's Great Expansion.

TQ-6 waits in the wings: Is there more to the issue than the numbers are being interpreted as meaning? In a reply, Nobel laureate economist Wassily Leontief [14] concedes some of Anderson's statistics but goes on to look at a number of cost thorns in his statistical rose garden.

True, the long recovery from the deep depression that brought President Reagan to power carried this country to the high point of the usual cyclical wave characterized by a low rate of unemployment and a high gross national product. It is more likely that wholesale tax cuts inaugurated by Mr. Reagan have made the level of the G.N.P., as measured by the

Government statisticians, several billion dollars higher than it would otherwise have been. But at what a cost!

Drastic cuts in public spending (except for military purposes) left the physical infrastructure of this country in ruin. City streets and transportation facilities, water supply and sewage systems, particularly in large metropolitan areas, are collapsing, the once glorious interstate highways are crumbling, and cramped airports are incapable of handling the rapidly increasing traffic. Despite the valiant effort of the under-financed, under-powered Environmental Protection Agency, our lakes, rivers and forests are succumbing to deadly acid rain.

What is even worse, the intellectual, cultural and social infrastructure of the country has suffered even more during this greater-than-ever boom than its physical counterpart. Primary and secondary schooling have been so weakened that a whole generation of boys and girls can hardly read, write or count, while the soaring price of higher education makes it impossible for many young people to take advantage of it.

No wonder the competitiveness of the United States is rapidly declining; many of our high technology industries are losing one battle after another in the struggle for their share of the foreign and even their own domestic market. At the same time, the rich are getting richer, and the poor are getting homeless.

Let us hope that contrary to Mr. Anderson's expectations the "Reagan Boom" will not continue in its present form for four or eight more years. If it does, the United States will find itself entering the 21st century as the richest country (in total value of stocks and bonds traded on the stock exchanges), but culturally and socially less advanced than other developed countries.

A Junk Poll?

The March 20–26, 1993 issue of *TV Guide* contains a survey by Ross Perot's Reform party, National Referendum–Government Reform, which readers are invited to respond to. Question 13, for example, reads: "Should laws be passed to eliminate all possibilities of special interests giving huge sums of money to candidates?"

Problems? For one, the results obtained from the poll are based on self-selected respondents and cannot be considered representative of a wider population group (see point 3). Also, Question 13 is a leading question, which is hardly surprising when the views expressed by Perot's Reform party are taken into account (see points 2 and 5). The term "special interests" as it is used carries with it the ominous suggestion of special interests taking over the country. When stated in this form in a Time/CNN poll, it received an approval rating of 80 percent. The Time/CNN poll also put the question as follows: "Should laws be passed to prohibit interest groups from contributing to campaigns, or do groups have a right to contribute to the candidates they support?" Forty percent of the Time/CNN respondents stated they would prohibit interest groups from contributing, while 55 percent stated that they had a right to contribute.

A Poll of AIDS Doctors Finds That Many Help in Suicides

This notice [15] refers to an article being published in *The New England Journal of Medicine* which reports that a survey of 118 members of the Bay Area Community Consortium, an association of local AIDS doctors, found that 53 percent of the respondents had reported helping at least one of their patients to commit suicide.

Further details about the poll are not given, which by itself should prompt us to raise a number of caution flags about the reliability of the 53 percent figure. When one considers that assisting suicide is illegal and may subject a practitioner to criminal and civil actions, there is good cause for one to question the reliability of the 53 percent figure.

STRENGTHENING THE SHIELD AGAINST BOGUS NUMBERS

Application of the aforenoted 11 test questions and poll points sets up a shield which, to a degree, can protect us from bogus numbers. We can strengthen this shield by means of the following measures.

Challenge Number Dropping

With number dropping having reached a feverish level in debates, presentations, and articles, one filter to help stem the flow of bogus numbers and their mongers is to publicly challenge the veracity of figures presented, as the occasion arises. The burden of proof of a figure's veracity should always be understood to be the responsibility of the presenter, not the presentee.

If this practice were implemented on a large scale, presenters of figures, data, and statistics would be more concerned about their veracity and bogus number mongers would lose their credibility through exposure.

Write a Letter Protest

When you encounter a newspaper or magazine article in which numbers are bandied about freely without justification or figures are cited without an appropriate context to make them meaningful, drop a short note to the editor; it need not be typed. This is an initiative all of us can take. One letter might not seem like much, but individual letters matter and they have a cumulative impact.

Call It What It Is

One obstacle to coming to grips with number lies is in the euphemisms that have been adopted to "soften" the harshness of the practice. Terms such as "reworking" the figures, "massaging" the numbers, carrying out a "management adjustment," and "creative accounting" are more innocuous sounding than falsifying numbers and figure fraud. The sugar coating is so thick at times that we are in danger of losing sight of what is in the pill, which, of course, is the idea. The bottom line is that figure fraud is figure fraud, no matter how much sweetener is applied, and it would make matters clearer for all concerned to get it out in the open by calling it what it is.

Promote Number Ethics

One of the most successful ways of reaching people is through an appeal to ethical behavior, which should include what might be called number ethics. Number, manipulation, exaggeration, distortion, creativity, call it what you will, is not a victimless practice. We are all, directly or indirectly, victims. Product safety personnel who participate in the suppression of data which show their product in an unfavorable safety light are collaborators in the destruction, misery, and death that their product subsequently unleashes because of safety flaws, as are the company executives who participate in the suppression. Government and political figures who manipulate data for a "higher good" or "cause" as they see it, or for personal political advantage, contribute to the undermining of our political, economic, and social institutions, set the stage for unrealistic decisions being taken, and open the door to the waste of an incalculable amount of resources.

Part Two

MATH MODEL PORTRAITS

INTRODUCTION

Mathematics, as we have observed, has come to be seen as the epitome of an ideal information state from which ambiguity, ignorance, and prejudice have been banished. One reason for this is the high respect accorded "mathematical proof," which is founded on the general acceptance of the following views.

4. Mathematical proof is the most reliable means for objectively establishing truth.
5. The precision of mathematical methods (that is, the techniques used to construct a math proof) guarantees unassailable conclusions that serve as pillars of stability and strength in a world besieged by foggy thinking, prejudice, and rampant special interests.
6. Mathematically derived conclusions are indisputable because they are based on deductive logic, which is untainted by bias and ideology.
7. Mathematical methods have the advantage of being certain. No qualified person can resist the truth of a mathematical conclusion properly communicated. The job of communication may be difficult if the solution is complex, but when the communication is competent, agreement is inevitable.

These views of mathematics express a state of affairs we should like to see realized in practice and which many of us believe has, to a high degree, been realized in practice. This makes us targets and potential victims of math manipulators who practice their ignoble trade for personal gain or in the interest of what they see as a "worthy cause." Many of us are also vulnerable to self-math-victimization because of our own misunderstanding of what math proof can do for us and its limitations.

To address the issue of what math-proof can and cannot do for us it is useful to introduce the term *math model*. A math model is a portrait in mathematical terms of any situation of interest to us. It consists of assumptions, called postulates, which the math artist uses as a starting point, and conclusions which are deduced from the postulates in a deductive-logical sense. These math-based conclusions are called valid consequences of the postulates. They are also called theorems and, depending on the context, are sometimes termed projections. The process of developing a math model is called math modeling.

Chapters 8 to 10 explore the nature of math modeling by considering the problem of developing a math model for the financial cost of smoking, looking at a situation involving a vacation trip, and examining the math model development process in general. Chapter 11 considers the interplay between assumptions and math based conclusions in public policy settings. Chapter 12 addresses the interplay between assumptions, math-based conclusions, and the selling of a point of view to the public at large. Chapter 13 returns to the aforenoted views on math proof armed with a perspective gained from Chapters 8 to 12. It seeks to help us develop self-defense mechanisms so that we can better evaluate arguments and points of view given respectability by math proof.

8

What's the Cost of Smoking?

For many years an artist friend of mine declined to act on the health warnings of his doctor about the consequences of smoking. But recently the doctor inquired as to whether he had any idea about how much it was costing him, and proceeded to note a few figures. My friend gave up smoking immediately.

A MATH MODEL FOR THE COST OF SMOKING

To be sure, not everyone is prepared to follow in my friend's footsteps, but I believe it would be informative to give attention to the cost of smoking from a math modeling perspective. Motivated by a desire to be "cool" and fit in with the crowd, young people sometimes take up smoking while in their teens.

Jules Warner considered the case of a teenager who takes up smoking at 15 and persists with the practice for 50 years. The price of a pack of cigarettes will vary over 50 years, but considering what the current price range of a pack is, suppose $4.00 is taken as an estimate of the average price of a pack over the next 50 years, Jules reflected. A new smoker may begin modestly with a few cigarettes a day, and then work up to a pack, 2 packs, and perhaps eventually 3 packs or more a day. Take a pack and a half as an estimate of the average amount smoked per day, over 50 years, Jules thought it reasonable to assume based on the data he had obtained. This yields a cost figure of 1.5 packs per day, times $4.00 per pack, times 365 days, equals $2190 per year, on average. Multiplying this figure by 50 yields $109,500 over a period of 50 years.

If our smoker gets married and has children he would probably give serious thought to obtaining life insurance, Jules considered. Insurance rates differ for smokers and nonsmokers, and what might cost a smoker $1000 a year might only be on the order of $700 a year for a nonsmoker, for a $300 difference, Jules' data suggested to him. If the insurance is held over 30 years, say, the additional cost for a smoker comes to $9000.

And then there are cleaning bills. If our smoker is not happy about his clothes, furniture, carpets, and draperies reeking of the smell of smoke and decides to do something about it, the extra cleaning bills might come to the order of $400 per year. Assume 30 years for this, too, Jules thought, and the cost comes to $12,000.

And then there is the matter of clean teeth. If our smoker is sensitive about yellow teeth, he might require at least one extra cleaning a year at, Jules thought it reasonable to assume, $90 a cleaning. If 30 years is taken for this too, we have $2700 as the total cost for teeth cleaning.

In summary, Jules's math model portrait for the cost of smoking consists of the following postulates and theorems.

Jules Warner's Math Model for the Cost of Smoking

Postulates

P1. A person takes up smoking at 15 and continues the practice for 50 years.

P2. The estimated average cost of a pack of cigarettes over the 50-year period is $4.00.

P3. The estimated average amount smoked over 50 years is a pack and a half a day.

P4. The additional cost of life insurance for a smoker is $300 a year. This applies over a period of 30 years.

P5. The additional cost of cleaning of clothes, furniture coverings, drapes, carpets, etc. for a smoker is, on average, $400 a year over 30 years.

P6. The additional cost of teeth cleaning for a smoker is, on average, $90 a year over 30 years.

Theorems

T1. The estimated cost of cigarettes over 50 years is $109,500.

T2. The estimated additional cost of life insurance over 30 years is $9000.

T3. The estimated additional cost of cleaning clothes, furniture coverings, drapes, carpets, etc. over 30 years is $12,000.

T4. The estimated additional cost of teeth cleaning over 30 years is $2700.

T5. The estimated total cost of smoking over 50 years is, rounded off, $133,000.

It all looks so precise, and in a deductive-logical sense, which defines the meaning of mathematical sense, it is; if you grant Jules's postulates P1 through P6 as a starting point, you must also grant his theorems T1 through T5 as being valid consequences of them. T1 follows from P1, P2, and P3; T2 follows from P4; T3 follows from P5; T4 follows from P6. The mathematical tool in each of these cases is our good old friend multiplication. T5 follows from T1 through T4 by another good old friend, addition. We should be careful not to allow familiarity to breed a take-for-granted attitude. As humble as these good old friends might strike us, they are as successful in getting the job done in the sense of validity as the most sophisticated cutting edge mathematical tools around today.

ADDITIONAL STRINGS ATTACHED

There are additional strings attached to this story which we should not overlook, and these are concerned with the features Jules left out. The $133,000 figure is in terms of today's dollars. Jules did not attempt to adjust for inflation over 50 years. He also did not consider other factors which may play a role in the cost of smoking, and perhaps this should be explicitly noted as postulate P.

P. *Other factors that may play a role in the cost of smoking are not being considered.*

It is not current practice to explicitly list P with the postulates of a math model, but we would be wise not to fall into the out-of-sight, out-of-mind frame of mind.

What other factors are not being considered in this model? In an article which prompted the refinement developed here, Hubert Herring notes some of the following [1]:

F1. In a study conducted by the National Bureau of Economic Research it was found that smokers earn 4 to 8 percent less than nonsmokers. As Herring points out, "this is a tricky statistic," and for this reason this dimension to the cost of smoking is not pursued here.

F2. Some 200,000 fires a year are started by smoking materials, which makes smokers more vulnerable to catastrophic loss.

F3. Expensive puffs. Some smokers have been known to take connecting flights so they could have a cigarette break at an airport, rather than taking a smoke-free cross-country flight.

F4. Smokers generally eat less than nonsmokers, which translates to a saving on food.

F5. Smokers are generally more vulnerable to a number of health-related problems, which translate to a cost. One dimension of this is seen in the difference in life insurance costs for smokers and nonsmokers.

DISAGREEMENT

On examining the postulates of Jules's model, Janet Wright, based on data she had obtained, was prompted to take issue with the realism of P2, P3, P5, and P6. Janet's data and study led her to introduce another math model for the cost of smoking by replacing P2, P3, P5, and P6 by P2a, P3a, P5a, and P6a, and introducing postulate P7 on health cost.

Janet Wright's Math Model for the Cost of Smoking

Postulates

P1. A person takes up smoking at 15 and continues the practice for 50 years.

P2a. The estimated average cost of a pack of cigarettes over the 50 year period is $6.00.

P3a. The estimated average amount smoked over 50 years is 2 packs a day.

P4. The additional cost of life insurance for a smoker is $300 a year. This applies over a period of 30 years.

P5a. The additional cost of cleaning of clothes, furniture coverings, drapes, carpets, etc. for a smoker is, on average, $600 a year over 30 years.

P6a. The additional cost of teeth cleaning for a smoker is, on average, $150 a year over 30 years.

P7. The additional cost of health maintenance for a smoker is, on average, $600 a year over 50 years.

Here, too, implicit in the formulation of Janet's postulates is the assumption that other factors which may play a role in the cost of smoking are not being considered.

Theorems

T1. The estimated cost of cigarettes over 50 years is $219,000.

T2. The estimated additional cost of life insurance over 30 years is $9000.

T3. The estimated additional cost of cleaning clothes, furniture coverings, drapes, carpets, etc. over 30 years is $18,000.

T4. The estimated additional cost of teeth cleaning over 30 years is $4500.

T5. The estimated additional cost of health maintenance over 50 years is $30,000.

T6. The estimated total cost of smoking over 50 years is, rounded off, $281,000.

Jules: The total cost of smoking over 50 years is $133,000.
Janet: The total cost of smoking over 50 years is $281,000.

Who's "right"? Could both be "wrong"? What does "right" mean? What does "wrong" mean?

WHO'S "RIGHT": JULES OR JANET?

Both results are right in the sense of validity; $133,000 is a valid consequence of Jules's postulates and $281,000 is a valid consequence of Janet's postulates. Jules's and Janet's reflections led them to different postulates, which in turn led to different valid conclusions. As to which, if either, valid conclusion is right in the sense of accurately describing reality, we are not in a position to say.

"PRECISE" MATHEMATICAL REASONING ACCORDING TO UNCLE GEORGE

As simple as they may be, the proofs of the cost of smoking theorems provide us with a view (partial view, to be sure) of what math proof can do. The tale of Uncle George provides us with a sense of what math proof cannot do.

While on his way to his Aunt Alice's birthday celebration, Jules Warner more and more began to feel that the powers that be were in conspiracy against him. It was snowing, traffic was bumper to bumper, and then someone in an old oldsmobile sideswiped his new Buick and took off. Jules finally arrived at his aunt's house, but there was no place to park and it took him 30 minutes to find a parking space which, alas, was two blocks away. Snow had turned to rain and Jules stepped out into a wicked downpour.

He finally made it to his aunt's house and his first thought was, I need a drink. But the powers that be were not finished with Jules. Between him and the bar stood a captive audience being "educated" by Uncle George. George firmly believed that he knew everything about everything and today his subject was mathematics, geometry in particular. "Since their truth was established by the precise mathematical reasoning for which the ancient Greek mathematicians are justly famous, the truth of the theorems of Euclidean geometry is beyond question," bellowed George in his most authoritative sounding tone.

Jules's attitude toward George was one of affection and, within limits, toleration of his education policy, but now George was really pushing it. He stood between him and the bar and he was talking nonsense in a voice which reflected the attitude the louder the pronouncement, the more profound the wisdom.

As usual, he got it all wrong, thought Jules. Precise mathematical reasoning establishes the validity of the theorems of Euclidean geometry, not their truth. Their truth is another issue, Jules further reflected. Do I want to straighten him out? Nothing would give me greater pleasure except, perhaps, that drink, and if I correct him I would start an argument that might go on forever and I'd never get that drink. Desire for the drink won and Jules made a run for the bar, waving and saying, "How are you, Uncle George, it's good to see you."

BECAUSE OF PRECISE MATHEMATICAL REASONING THE TRUTH OF EUCLIDEAN GEOMETRY IS BEYOND QUESTION.
SOUNDS GOOD AFTER A DRINK!
SOUNDS BETTER AFTER TWO DRINKS!
SOUNDS LIKE NONSENSE BEFORE A DRINK!

9

Andy and Rasa Take Vacations

ANDY PLANS A TRIP

Recently, Andy was engaged in planning a car trip from home in Brooklyn, New York, to the popular vacation town of Kennebunkport, Maine, in mid-August. His problem was to set up a math model for the trip which would enable him to predict the total time required for the journey.

The setting of any such problem presents numerous features and characteristics, many of which are irrelevant or unessential to the focus of the problem. In developing his math model, Andy had to sort this out and decide on which features were fundamental and which were negligible. This required discretion and judgment, the most controversial aspect of the math model development process; one person's essential might be another's irrelevancy.

Andy examined a map and laid out a route. Based on data provided by friends who had recently made the trip, he made assumptions about departure time, weather conditions, the traffic flow to be expected along various points, speeds that would be possible, and the number of rest stops to be made and their duration. Such considerations led him to a math model consisting of a line segment 330 miles long joining points representing Brooklyn and Kennebunkport, and the problem of determining how long it would take an object moving at an average speed of 55 miles per hour to cover this distance.

Andy's math model is an idealized, abstract rendering of the real situation involving a trip from Brooklyn to Kennebunkport. It is intended to capture the main features involved in taking such a trip and reflects these features as he sees them and the assumptions that he was led to make. As in the case of Jules vs. Janet concerning the cost of smoking, it is possible that someone else planning such a trip would see things in another light and compose a very different math model.

By employing division, we obtain the valid conclusion that an object moving along the idealized path of Andy's model at an average speed of 55 miles per hour would take $330/55 = 6$ hours to make the journey.

In summary, Andy's math model consists of the following:

Andy's Math Model

Andy's postulates

P1: A line segment of length 330 miles joining points representing Brooklyn and Kennbunkport is taken as an idealized representation for the actual path of the trip.

P2: A point moving at an average speed of 55 miles per hour from the point representing Brooklyn to the point representing Kennebunkport along the line segment is taken as an idealized representation of the car trip itself.

Andy's theorem

T1: The time required to make the trip is 6 hours.

Andy's math model is a math portrait of his journey. It obviously is not a photographic likeness, but as with people portraits, which may vary considerably depending on the artist, we are prompted to inquire: Is it accurate?

One grip on this question is obtained through Andy's theorem. The acid test involves undertaking the journey, noting the time required, and comparing it with the projected time of 6 hours from Andy's theorem. If there is a "small" discrepancy between the actual and projected times, this would establish that Andy's theorem is realistic in this case and, by reflection, be evidence in support of the realism of his math model portrait of the journey. If there is a "large" discrepancy between the actual and projected times, this would establish that Andy's theorem is not realistic in this case and, reflecting back, lead us to conclude that his model is not a realistic portrait of the journey at hand.

The status of Andy's theorem as a valid conclusion derived from his postulates is not at stake here. Division, yielding $330 \div 55 = 6$, establishes the validity of Andy's conclusion and confers on it the status of theorem. To make a judgment on whether a valid conclusion is realistic or not, we must turn to the real world. In Andy's case, it reduces to taking the trip and observing the outcome.

Andy took his vacation as planned in the middle of August and it took him 5 hours and 55 minutes, which establishes his theorem as realistic in this case and provides evidence in support of the realism of his math model portrait for the trip.

RASA'S TRIP

Andy's sister Rasa was planning to take a brief vacation trip to Kennebunkport during Labor Day weekend. Andy's math model worked well for him and she decided to follow the route it prescribed, expecting the journey to take around 6 hours.

Rasa took the trip on Labor Day weekend as planned, but it took her 6 hours and 50 minutes. This actual trip time differs considerably from the projected 6-hour trip time of Andy's theorem, so that something clearly went wrong; but what? "Your theorem stinks," Rasa shouted at her brother in a somewhat agitated manner. "It's wrong, it's not valid," she continued.

Rasa's experience proved Andy's theorem wrong in terms of reality, but not validity. Andy's theorem is a false statement as a description of the travel time to Kennebunkport on a Labor Day weekend, but it remains a theorem. It is still an inescapable consequence of Andy's model—more specifically, his postulates, which is what makes a theorem a theorem. "Math proof," division in this case, did its job in yielding a valid conclusion. While validity is not the same as reality, they are linked in this way:

If the postulates of a math model are realistic, so will be the theorems obtained from them. If a theorem is found to be false, this sends us a signal that some of the postulates of the math model are unrealistic and require modification.

Andy's model, Rasa's experience showed, is not realistic for travel to Kennebunkport on a Labor Day weekend. In reexamining Andy's postulates, we find that they do not realistically take into account unusually heavy traffic delays, characteristic of holiday weekends, around the tollgates of the Whitestone Bridge. Further examination of Rasa's actual trip shows that this is where she had the difficulty.

Rasa undertook to modify Andy's model to make it more realistic for travel to Kennebunkport on a Labor Day weekend. She reviewed Andy's assumptions and took into account data on traffic delays around the tollgates of the Whitestone Bridge on holiday weekends.

Rasa's Modification of Andy's Math Model

Rasa's postulates

P1: A line segment of length 330 miles joining points representing Brooklyn and Kennbunkport is taken as an idealized representation for the actual path of the trip.

P2: A point moving at an average speed of 49 miles per hour from the point representing Brooklyn to the point representing

Kennebunkport along the line segment is taken as an idealized representation of the car trip itself.

Rasa's theorem

T1: The time required to make the trip is 6 hours and 44 minutes.

Rasa's theorem is in close agreement with her experience, which is evidence in favor of her math model as a realistic portrait of travel from Brooklyn to Kennebunkport on Labor Day weekends under current conditions.

NINO VS. RIRO

Whether Rasa's math model will serve as a suitable travel portrait in the future depends on the extent to which "current conditions" are maintained. Andy's model and Rasa's modified version of it took into account traffic delays due to road repairs that were in progress. When these repairs are completed, current conditions will have changed in a significant way and a suitable modification of both Andy's and Rasa's models would be in order to keep them current.

As we have seen, one way to get a grip on the question of a math model's realism is through its theorems. The other way is through its assumptions, formulated as postulates. Are the assumptions realistic? We should always keep in mind that in addition to the explicitly stated assumptions there is the implicit assumption that the factors not explicitly addressed by the assumptions are not being addressed or are being viewed as negligible to the focus of the situation under study. If the answer is *yes*, the assumptions are realistic, and this is a correct assessment, then the theorems will be realistic as well. The RIRO principle operates; if realistic input, then realistic output in terms of theorems. If the answer is *no* to at least some of the assumptions, and this is a correct assessment, then we cannot be sure about the realism of the theorems. Some might be right on target, whereas others might be considerably off reality's mark. The NINO principle operates; if nonsense in, there is a good chance of nonsense out, even though some theorems might be realistic.

Rasa was caught by the NINO machine. She took over Andy's math model without carefully examining its assumptions and paid the price in terms of an unrealistic theorem about travel time.

10

Math Model Development

THE MATH MODEL BUILDING PROCESS

The development of math models for the cost of smoking and Andy and Rasa's math model experiences reflect in miniature the general nature of math model building. It is useful to identify in general terms the factors involved in math modeling.

The development of a math model for a situation consists of the following steps.

1. *Specify the situation (process or problem) to be studied.*
2. *Collect data about the situation, make "suitable" assumptions about the factors involved, and formulate an idealized representation or portrait, if you will, for it.*

This is the hypothesis of the math model, and the individual statements that make it up are its postulates. They are based on the data collected and assumptions made from studying the situation.

Reflecting on Andy's development of a model for his trip, we see that he collected data about distances of various parts of the route he intended to take, travel times, and made assumptions concerning departure time, weather conditions, number of rest stops and their duration, and a number of other factors. Andy's postulates are a crystallization of these data and assumptions.

3. *Apply "math proof" to obtain valid conclusions, the theorems, with respect to the hypothesis of the model.*

In principle, it matters not whether the math methods employed to carry out math proof are simple, as in the cost of smoking and Andy's trip situations, or the ultimate in technical sophistication.

4. *Test the accuracy of the model as a portrait of the situation under study by comparing what the theorems say about the behavior of the situation with its actual real-world behavior.*

"Close" agreement between what the theorems say about reality's behavior and reality itself is viewed as evidence in support of the math model as a realistic, though idealized, portrait of the situation. It does not establish that the model is in any sense a "perfect" description. Just as a person's portrait may be sketched in many ways, so too may a math model portrait be developed in many ways. The math artist must exercise judgment and discretion as to which features of the situation are of paramount importance and must be taken into account in the portrait developed in the model, and which features are negligible and need not be reflected by the model. This calls for insight and sensitivity which, needless to say, cannot be pinned down in a sharp, unequivocal fashion. We

can expect different math models to emerge from math artists who see a situation in different terms.

Suppose two or more math models become available for a situation. How do we choose between them? The decisive verdict is rendered by the behavior of reality. The fundamental question is: Are the statements

made by a math model's theorems in close agreement with the behavior of realty? When a math model theorem is found to be in disagreement with reality, this tells us that some part of the model's hypothesis is incomplete or unrealistic as a portrait of the situation at hand.

A theorem found to be false does not lose its status as a theorem. It remains a valid consequence of the hypothesis of the model, correct in the sense of validity, incorrect in the sense of reality.

5. *Refine the math model portrait of the situation under study when observations of the situation's actual behavior differ "significantly" from the behavior predicted by the model's theorems.*

Rasa's experience with the travel time theorem of Andy's model led her to modify his model to make it realistic for Labor Day weekend travel from Brooklyn to Kennebunkport.

Figure 10.1 summarizes the basic steps in the development of a math model in diagrammatic form.

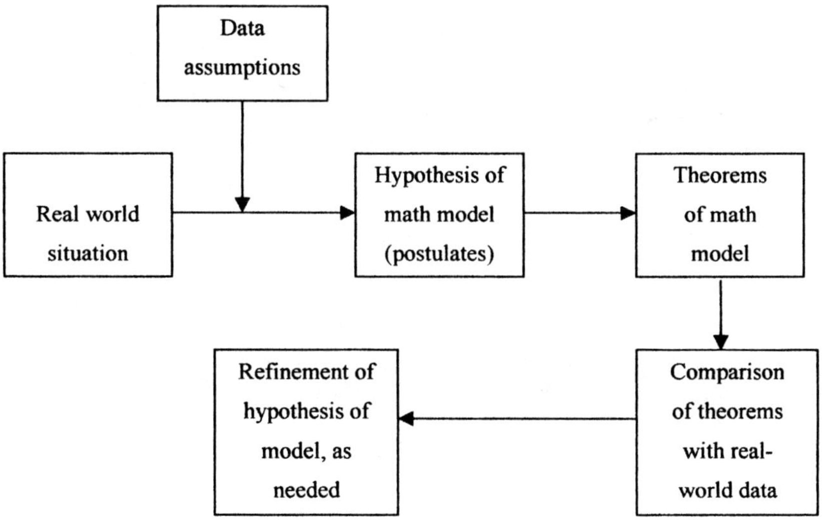

Figure 10.1 Development of a math model.

A SECOND DIMENSION OF MATH MODELING

Another dimension of math modeling arises when the theorems obtained from the model either cannot be compared with the behavior of reality or doing so might have dire consequences. The Axel Company, a producer of television sets, is eager to introduce two new models, TV-1 and TV-2, into the market. A theorem of a math model M1 developed for the company by its operations research department says that a monthly production schedule of 500 TV-1 and 600 TV-2 units would generate the largest possible monthly profit of $300,000. One way to determine if this theorem is realistic is to implement the production schedule and see what happens. If what happens is that a substantial monthly loss is incurred, the Axel Company could be forced out of business. Understandably, it is hesitant to take the plunge.

The alternative is to carefully, very carefully, examine the hypothesis of the production model. If it is indeed agreed that it is realistic, and this is a correct assessment, RIRO (if realistic input, realistic output) will prevail and it would make sense to implement the production schedule. If there are reservations about the realism of the assumptions, beware of NINO (if nonsense in, nonsense out). This is the best advice that can be followed in general.

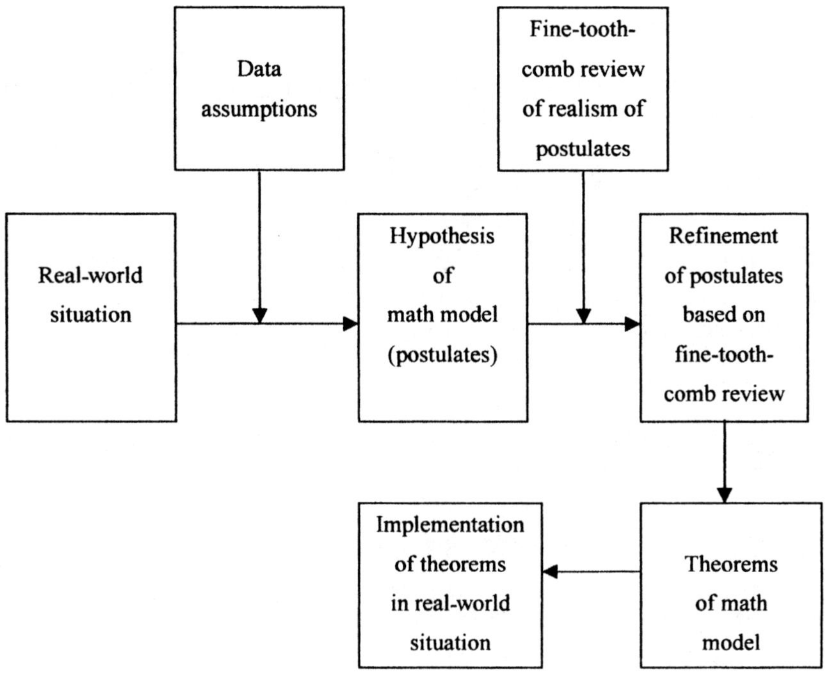

Figure 10.2 Development and implementation of a math model.

Figure 10.2 summarizes the basic steps in the development and implementation of math models in situations where implementation of unrealistic theorems could be disastrous.

WHICH SOLUTION SHOULD BE IMPLEMENTED?

In the spirit of let's minimize risk by covering our bets, the Axel Company hired an outside consulting firm to independently study the TV production situation and make recommendations. Its analysis of the data collected led the firm to formulate a math model, M2, from which they deduced the theorem that to obtain the maximum monthly profit, $220,000, the monthly production schedule should be set at 400 TV-1 and 700 TV-2 units.

Now what? It's always easier when you have a choice of one; take it, or leave it. But a choice of two is another matter. Bottom-line Bob, chairman of the 10-member board charged with implementing the "best" production schedule, argued that it's obvious what should be done. "Implementation of M1 brings us a monthly profit of $300,000, whereas implementation of M2 brings us a monthly profit of $220,000. It's a no brainer; we should go with M1." The board voted 9 to 1 to implement M1.

Alas, instead of a $300,000 monthly profit being realized after M1 was implemented, a $50,000 monthly loss was incurred, and two years later the Axel Company's new venture into the television market had to be written off. Bottom-line Bob, who received credit for this disaster, was confused, upset, and out of a job. He went to Reflective Ramunė, the chair of the new board and the one person who had voted against implementation of M1, with some questions.

"PRECISE" MATHEMATICAL REASONING ACCORDING TO BOTTOM-LINE BOB

"Ramunė, I don't understand this. I always liked math in high school and college. Solving equations, factoring expressions, differentiating functions, throwing the data into the computer and letting it do its thing—I got a real kick out of that. What I like most about math is its precision. You don't get 10 sides to a story. You get one answer and that's that; no baloney. Give me the numbers and that ends all arguments; no more beating around the bush."

"Does this mean there's no room in your number world for others who might seek to make sense of the numbers?" inquired Ramunė.

"What's there to make sense of?" replied Bob. "Numbers speak for themselves."

"But through whose judgment, and how reliable is that judgment?" countered Ramunė.

"My judgment," replied Bob, "that's what Bottom-Line means."

"Yes, but I'm afraid that that may be quicksand masquerading as bedrock," answered Ramunė.

"It turned out to be quicksand in this situation, which is why you're in the predicament you now find yourself."

"Bob, I think your math courses may have focused too much on technique and not enough on perspective. Techniques can be fun to a point, but without perspective on its place in the overall role of mathematics in applications we see only a small tip of the mathematical iceberg.

Mathematics is precise in the sense that it gives us valid conclusions based on the assumptions made, which is where technique—factoring, solving equations, differentiating functions, and the like—plays its major role. Whether the assumptions made are realistic or not is a matter which technique can't help us with. The question of how to formulate these assumptions and reach a judgment on their realism may indeed yield ten sides to the story. I'm afraid that those who find mathematics attractive because of what they perceive to be its absolutist nature have misunderstood the meaning of mathematical precision."

"Model M1's profit projection was $80,000 more than the projection of M2. Why did you vote against implementing M1?"

"It all comes down to the assumptions, Bob. Since I had reservations about the realism of M1's assumptions, I had no faith in the realism of its profit projection."

BUT I USED A COMPUTER!

"Ramunė, I still don't quite get it. The company spent millions of dollars to update its computer system. I had access to the latest and the best. Why didn't this guarantee success?"

"The statesman Henry Clay's observation that 'statistics are no substitute for judgment' applies equally well to the computer, Bob. We cannot expect the computer to convert unrealistic assumptions into golden truths. Keep in mind NINO: If nonsense in, nonsense out. I'm sorry to have to say that indiscriminate use of computer power has made possible the generation of more nonsense from unrealistic assumptions more quickly than ever before by people having little understanding of what they are doing."

"But then, what good is this super computer technology to us?"

"For number crunching and delivering results quickly and efficiently, the computer is without equal. A symbiotic relationship has developed between the math model building process and computers. Computers have made it possible for us to solve previously unapproachable large-scale problems that come out of math models, while the accessibility of such problems to computer solution has made possible the use of complex math models. But none of this overrides NINO."

11

As Ye Sow, So Shall Ye Reap

The three most important considerations in buying real estate, an old saw has it, are location, location, and location. Much the same may be said about developing a math model.

Just replace location by assumptions; for I'll "prove" whatever you want, if you'll grant me my assumptions. If you refuse to do so, I, like Superman in the presence of Kryptonite, will be powerless.

BASE CLOSINGS

By the early 1990s there was general agreement that some military installations should be closed as a cost-cutting measure. The question is, which ones? Maybe yours, but not mine, was the answer offered by local politicians as the struggle to save local bases heated up.

The Rome Laboratory in Rome, New York, a high-tech Air Force research installation, came under scrutiny. An analysis carried out in October 1994 of a proposal to close the lab and move its facilities to Hanscom Air Force base in Massachusetts led to the conclusion that it would cost \$133.8 million to move the lab with annual savings of \$1.5 million being realized. It would take 90 years to recover the cost of the move at this rate. The Pentagon did not recommend closing the Rome lab.

In February 1995 the Air Force released an analysis of a proposal to close the Rome lab and move 60 percent of its operations to Hanscom and 10 percent to the Army base in Fort Monmouth, New Jersey. It led to the conclusion that relocating the lab would cost \$52.8 million with annual savings of \$11.5 million. The cost of the move would be recovered in five years at this rate. The Pentagon recommended closing the Rome lab.

In May 1995 another analysis led to the conclusion that relocating the lab as described would cost $79.2 million with annual savings of $13 million. The cost of the move would be recovered in six years at this rate. Again the Pentagon recommended closure.

Which numbers, if any, are realistic? They arise from different assumptions, which means that the question of number realism falls back on assumption realism and the question; Which assumptions would you be willing to grant?

In June 1995 a presidential commission reversed the Pentagon's closure recommendation. It gave a 13-year estimate for the time it would take to recoup the cost of relocating the lab, which it was felt was too long to justify its closure. Is 13 years more realistic than the previous estimates of 90, 5, and 6 years? Maybe, maybe not, but again we are taken back to the assumptions.

IDEOLOGY VS. ECONOMIC REALITY

As President Jimmy Carter was preparing to leave office in January 1981, the first priority of the new Reagan administration was to conduct a thorough overhaul of the Carter budget for fiscal 1982, which was to begin on October 1, 1981. David Stockman was appointed Director of the Office of Management and Budget and Reagan's budget team set to work in January 1981.

Stockman notes [1]: "There were three doctrines represented on the forecasting team: the monetarists, the supply-siders, and the eclectics. . . . The new chairman of the Council of Economic Advisors, Murray Weidenbaum, tended toward the third approach."

In conversation with the journalist Laurence Barrett, Weidenbaum [2] commented that: "It was a forced marriage. The supply-side people insisted on [forecasting] rapid growth in real terms and the monetarists insisted on rapid progress in bringing down inflation. Each of them would go along with a set of numbers as long as their own concern was satisfied. The monetarists weren't that concerned about growth and the supply-siders weren't that concerned about inflation."

Weidenbaum was a latecomer to the initial negotiations on economic assumptions. Barrett [2] reports that "he was so shocked at what he found that he seriously considered resigning even before unpacking the cartons in his new office. He had the same urge a few months later, when the assumptions were reviewed and retained for political reasons, though by then everyone knew they were specious." Weidenbaum stayed until the summer of 1982 because he felt that, bad as things were, the economic assumptions would have been more irresponsible than they were had he not been there.

Stockman [1] notes: "The table that follows [Table 11.1] tells the whole story, proving that our Rosetta stone was a fake. . . . The February 1981 economic forecast eventually became known as 'Rosy Scenario.' Weidenbaum wrote the final specific numbers. But its underlying architecture—the push-pull hypothesis—was ultimately the work of a small band of ideologues."

Table 11.1 Real GNP Growth

Quarter	Supply-side monetarist consensus (%)	Final Weidenbaum forecast (%)	Actual outcome (%)
1981:4	4.0	4.0	−5.3
1982:1	9.4	5.2	−5.5
1982:2	7.8	5.2	0.9
1982:3	6.8	5.2	−1.0
1982:4	5.4	5.2	−1.3

Source: Ref. 1.

I PREFER "MATHEMATICAL SIMPLICITY," BUT . . .

The advertising media selection problem is to choose from various advertising media a selection which is, in some sense, best. Potential advertising media include newspapers, magazines, radio, television, and direct mailing; and then there are a number of options within these media. In most variations of this problem the objective is to maximize audience exposure. Budget is one of the constraints on the media selection to be made. Other restrictions on the kind of media mix selected might come from company policy, contract requirements, and limited availability of media.

In 1961 the firm of Batten, Barton, Durstine, and Osborn (BBD&O) came out with a math model called a linear program model (see "From Politically Incorrect to Nobel Prize," Chapter 4) for the media selection problem [3], which was followed by the publication of a number of articles on the application of linear program models to this problem. In the early 1960s hopes ran high in the world of advertising for linear programming, as the study and application of linear program models had come to be called, and in a spasm of optimism BBD&O placed full-page ads reading, "Linear programming showed one BBD&O client how to get $1.67 worth of effective advertising for every dollar in his budget."

As complex as they may sometimes be, linear program models have the desirable feature of being "mathematically simple" compared to alternatives that might present themselves. But are they realistic? The best that could be said in general is sometimes yes, sometimes no; it depends.

Initial optimism for linear programming turned to disappointment, which in turn gave way to a calmer understanding of the potential and limitations of this approach to media selection. This attitude transition is reflected in the examination, in a series of papers, of the realism of the assumptions that had served as a foundation for the initial linear programming approach to media selection. Philip Kotler [4] noted the following limitations:

1. Linear programming assumes that repeat exposures have the same effect.
2. It assumes constant media costs (no discounts).
3. It cannot handle the problem of audience duplication.
4. It says nothing about when ads should be scheduled.

Frank Bass and Ronald Lonsdale [5] explored the operational details of the application of linear programming to media selection and found linear program models to be "crude devices to apply to the media selection problem. . . . The linearity assumption itself is the source of much of the difficulty. Justifying an assumption of linear response to advertising exposures on theoretical grounds would be difficult. . . . Assumptions about the nature of response to advertising cause most difficulties in models of the type examined in this article."

In the final analysis efforts to employ linear programming to realistically handle sophisticated versions of the media selection problem were unsuccessful. This prompted the abandonment of linear programming for media selection.

LET'S ASSUME WE HAVE A CAN OPENER

An engineer, a chemist, and an economist are marooned on an island and they find a can of tuna, but have no way to open it. Says the engineer: "Let's find a rock and crush the can open." Suggests the chemist: "That's impractical: A better way is to find some chemicals and blast it open." Quoth the economist: "You people are truly misguided. There is only one way: Let's assume we have a can opener. . . ."

Gene Epstein, columnist for *Barron's*, reprises this tale to make an important point about commonly made economic assumptions [6]. Specifically, Epstein considers the article "Labour Allocation in a Cooperative Enterprise" by Amartya Sen, who a week earlier had been announced as the 1998 winner of the Nobel Prize for Economics. Epstein notes that Sen sets down assumptions which, he assures us, are "serious but not especially odd in this branch of economics." They include "well-behaved utility and production functions; automatic fulfillment of the second order conditions of welfare maximization and of equilibrium; no uncertainty; perfectly competitive markets; homogeneity of labor . . .—a whole set of can openers" whose realism, Epstein points out, is "truly dizzying for the colossal naivete they reveal." More generally, "Your typical economist," Epstein further observes, "assumes such absurdities as 'no uncertainty' in markets, even though uncertainty is any market's middle name, and 'well-behaved utility and production functions' (read: unchanging consumer tastes and production processes), even though such good behavior is the rare exception rather than the rule—and he assumes so for only one reason: Because that way, he gets to use a lot of math" [6]. (Linear programming applied to the advertising media selection problem, discussed in the previous section, is a good example of Epstein's general point.)

Epstein, as most of us, is not opposed to using a lot of math provided that the math conditions are suitable fits to the economic situation. Unrealistically forcing an economic structure, or any other for that matter, into a mold that fits the assumptions of a nice bit of math is as uncomfortable and unproductive as forcing a pair of size 6 shoes on size 9 feet, Epstein is essentially warning us. Economists do this all the time, he further points out.

This has become of increasing concern to economists and should be of major concern to the rest of us since we will have to cope with the devastating consequences of NINO. In the essay "The Future of Economics," *The Economist* puts it cogently: "If models are to reveal anything, they must be simpler than reality: the challenge is to simplify usefully" [7].

COMING TO GRIPS WITH PRODUCTIVITY

How well is the economy doing? One measure is provided by the productivity growth rate. Productivity measures the value of what a worker produces in goods or services during a given time period on the job. When output rises, so does the productivity growth rate. If Arnold Bloom's work brings his company an hourly revenue of $100 an hour, let us say, which then goes to $103 an hour, then Arnold's productivity growth rate is 3 percent (per hour). The additional revenue generated can go toward profit or salary raises or both. The productivity growth rate of the nation's work force as a whole is viewed as the path toward higher income and standard of living.

The numbers tell us that the 1950s and 1960s had productivity growing at nearly 3 percent a year, dropping to about 1 percent a year in the early 1970s, staying around that level until the mid 1990s, and beginning to accelerate around 1996. How secure are these numbers? They are tied to assumptions and if we accept the assumptions, the numbers follow, no ifs, ands, or buts. But more and more people whose opinions cannot be ignored have found reason to take issue with these assumptions.

During the late nineteenth and early twentieth centuries the U.S. economy rested primarily on agriculture. The new technology of that period centered on the internal combustion engine and electrification, whose benefits were not immediately reflected by the official productivity figures. From 1889 to 1929 farm productivity increased at an average annual rate of 0.9 percent, whereas manufacturing productivity experienced an average annual growth rate of 2.4 percent during the same period. Much the same thing has been happening in recent years with the introduction of computers and information technology. As Robert Solow, Nobel Laureate in Economics, observed a few years back: "You can see the computer age everywhere these days except in the productivity statistics." [8].

On November 12, 1999 the Bureau of Labor Statistics released an upward revision of productivity figures based on modified assumptions which better reflected the integration of computers, information technology, and enhanced financial services into the economy. Average annual productivity growth in the 1990s, determined as 1.5 percent with respect to the assumptions of the earlier productivity model, was now put at 2 percent with respect to the modified assumptions. The productivity slowdown of the 1970s looks less severe with respect to the modified assumptions than the old. For further discussion see Refs. 9 to 14.

COMING TO GRIPS WITH INFLATION

The consumer price index (CPI), a value constructed by the Bureau of Labor Statistics (BLS), is intended to measure the effect of price changes on a collection of some 400 goods and services, called a market basket, believed to make up the "typical" purchases of urban wage and clerical worker families and individuals, including professional, managerial, and technical workers, the self-employed, short-term workers, the unemployed, and retirees.

The CPI is a national index in that it serves as the nation's main measure of consumer inflation. It is the valid consequence of a highly sophisticated math model that employs statistical methods. To accomplish this the BLS tracks the prices of about 80,000 goods and services which are classified into six major groups and subgroups: food, housing, apparel, transportation, health and recreation, and other goods and services. It calculates how the items in these categories should be weighed by determining the relative amounts that consumers spend for them. Such inputs plus a number of assumptions are the ingredients which make up the hypothesis of what may be termed the CPI math model. The CPI value generated as a valid consequence of the hypothesis of this model is published in the third week of each month by the BLS. The monthly increments are added to yield an annual CPI figure. For comparison purposes the CPI is taken as 100 for the period 1982–1984, called the base period. At the beginning of 1995 the CPI stood at 155.3, which says that overall the prices of consumer goods and services (defined by the market basket) increased an average of 55.3 percent over the base period 1982–1984. At the beginning of 1996 the CPI stood at 158.6, a rise of 3.3 percent over 1995; the inflation rate increased by 3.3 percent over 1995, this says.

In addition to serving as a measure of the change in prices of goods and services, the CPI is used to define the purchasing power of the dollar, determine "real" disposable income, and to establish "cost-of-living" increases. At the beginning of 1996 the purchasing power of the dollar was 63.1 cents, relative to the 1982–1984 base period.

How solid are the numbers 158.6 and 0.631? As to being valid conclusions of a CPI math model formulated by economists and statisticians of the BLS, no question about it; as to being realistic measures of inflation, this is another matter which received a good deal of public attention in the mid-1990s. (See "The Consumer Price Index Must Be Reformed!" Chapter 12.)

Many people expect the CPI to reflect an individual's experience with inflation and become disillusioned when their experience differs considerably from what the CPI says. We should keep in mind that the

CPI is intended to reflect the purchasing experience of a population of people. An individual's inflation rate is tied to his or her own purchasing pattern, which may vary considerably from the average for a population.

WHAT'S THE "BEST" MATH MODEL?

We are often prompted to reflect on the "best" of a kind—"best" car, college, lawyer, health plan—best in the sense that it's top of the line and cannot be improved on. Considering the wide-ranging effect of the CPI on the many dimensions of our lives (see "Ripples," Chapter 15) it makes good sense to inquire about the "best" CPI model for inflation.

We grappled with the issue of "best" in "What's the "Best" Measure of the Iron Horse?" (Chapter 5) and we would do well to return to this discussion to help us get a grip on this issue here. What emerges is the realization that the hope for a best measure of the Iron Horse is untenable. The same considerations confront us in any attempt to define a best math model for whatever corner of reality happens to be under study, whether

it be a CPI model for inflation or any other. We have no choice, therefore, but to conclude that the hope for a best math model for reality is untenable and has to be abandoned.

What is feasible is for us to work on refining a math model to make it more realistic for our needs. (A good, simple example to keep in mind is Rasa's refinement of Andy's model for her Labor Day weekend trip to Kennebunkport; see "Rasa's Trip," Chapter 9.)

Which refinement of a math model would be seen as "suitable" involves a trade-off between degree of accuracy required, level of simplicity desired, and, if present, political considerations. (Concerning political considerations, see, for example, "NAFTA Will Increase Jobs," "The Consumer Price Index Must Be Reformed!," and "Social Security Is Collapsing," Chapter 12.)

The following insightful analogy about a "suitable" refinement of a math model description considering its purpose is due to Sir Rudolph Peierls [15]:

> If we look at the photograph of a landscape in snow, showing part of a snowfield in sunlight and another part in the shade, we would describe this photograph as having one area in very light shade of grey, almost

white, and another uniform area of a slightly darker shade of grey. On a more careful study of the same photograph under a microscope we discover that the grey areas are in fact made up of small black dots, the silver grains of the photographic process, on a white background, the darker area differing from the lighter by having more of the black dots.

In a sense this discovery has proved the first description wrong, but it would be more reasonable to say that the new description refines the old one and replaces it when we are concerned with far greater detail than was the case at first. The old description is still good enough when we are concerned with taking a photograph or with looking at it. In fact by being simpler it is more valuable for that purpose. If we had to think of photographs always as collections of black dots on a white background we should find the photographs quite useless as pictures.

THE HAZARDS OF LONG-RANGE FORECASTS

In December 1995 budget negotiators were quarreling over the size of the federal budget deficit anticipated for 2002. The Republican congress accepted the Congressional Budget Office's (CBO's) projected deficit figure of $340 billion which, in part, was based on the assumptions that national output would grow 2.3 percent a year and that no changes would be made in federal government programs and taxes.

The Clinton administration's Office of Management and Budget (OMB) projected a deficit of $216 billion, $124 billion less than the CBO's, assuming that output would grow 2.5 percent a year. The CBO assumed a slightly higher interest rate on the national debt than the OMB and faster growth in spending on Medicare and Medicaid (9.7 percent a year vs. 9.1% a year for the OMB).

Considering that the argument in 2000 was over how large the surplus will be in 2002, and the argument in 2002 is over how large the deficit will be and for how long, the obvious question is: How could so many, well-trained, "smart" people be so far off reality's mark? There are two factors that make long-term forecasting particularly perilous. First, even if the assumptions made are only slightly off reality's mark, what initially is only slightly off may be magnified over time into a very sizable error. And then, the longer the span of time involved, the greater the chance that troublesome, wholly unanticipated factors might arise to throw everything way off (the terrorist attack on the World Trade Center in New York on September 11, 2001, for example).

The response to these difficulties does not lie in throwing up one's hands and abandoning long-term forecasting as too unreliable to be useful, but in putting them in perspective as best guesses with respect to the existing information picture to be used for initial planning purposes. These best guesses should be updated as the existing information picture is refined, but still viewed as best guesses, although hopefully better guesses.

12

Hyperbole and Hype

THE ART OF MAKING HORSE FEATHERS LOOK GOOD

An interest group favoring a public policy position is faced by the problem of investing it with a sense of respectability and orchestrating the media, lawmakers, and the general public in its favor. This may be accomplished by a successful blending of two fundamental spin principles with an appeal to the idea of mathematical precision, and media orchestration.

Fundamental spin principle 1 holds that a position persistently repeated eventually gains acceptance as wisdom. Number 2 notes that a spade should never be simply called a spade; call it an indispensable agricultural implement, for example. The implementation of these principles rests on three components:

KISS: Keep It Simple Stupid.

MATH: State the position in numerical form to invest it with a sense of respectability through its association with the precision of mathematics.

BOLÉRO: Cultivate influential public figure and media endorsements to develop a sense of increasing intensity and momentum for the position, analogous to the effect achieved by Maurice Ravel in his composition Boléro.

BALANCE THE BUDGET

"Balance the federal budget" was taken up by the Republican leadership in the mid 1990s as a political war cry to strike at what they viewed as excessive spending for social programs and present President Bill Clinton with a perilous political problem. The sound of budget in balance satisfies the KISS and MATH requirements of the fundamental spin program and has a captivating ring to it, especially when misleadingly equated to the problem of balancing our own personal budgets. Due to BOLÉRO this war cry has achieved wide public acceptance. President Clinton sidestepped the thorny political problem thrust at him by adopting the budget-in-balance mantra as his own.

While budget in balance has a satisfying precise mathematical sound to it, when it comes to economic reality a substantial body of economic opinion holds that the economic health of the nation is much more realistically described by the ratio of the budget deficit to gross domestic product (GDP). To enshrine budget in balance as the condition to be given first priority misses the economic mark and could precipitate economic disaster should a recession occur, this view holds.

The triumph of budget in balance over deficit divided by GDP as the primary focus signals the political victory of KISS, MATH, and BOLÉRO over economic reality. A proposed constitutional amendment to balance the budget was turned back in the Senate in 1997 (by one vote), 1996 (three votes), 1995 (one vote), 1994 (four votes), and 1986 (one vote). The Senate passed it in 1982, but it lost in the House, as it did in 1990 (seven votes), 1992 (nine votes), and 1994 (twelve votes); it was passed by the House in 1995.

As Freddie of horror movie fame would say, I'll be back!

NAFTA WILL INCREASE JOBS

The early 1990s saw a contentious debate arise over the North American Free Trade Agreement (NAFTA), a five-volume, 15-pound plan that would link the economies of Mexico, the United States, and Canada. President George Bush pushed hard in support of the treaty and after the 1992 election President Clinton took up the bullhorn to secure congressional approval and public acceptance.

One dimension of spin principle 1 took the form of ads such as the full-page infomercial taken out by Citibank, titled "WHY 300 ECONOMISTS, 136 NEWSPAPER EDITORS AND 6 PRESIDENTS SUPPORT NAFTA: NAFTA WORKS FOR EVERYONE" [1]. Five claims were made in support of NAFTA, one of which commented on the highly sensitive question of whether America would gain or lose jobs if NAFTA were approved. "NAFTA will increase employment in the U.S.," it proclaimed.

No justification for this claim was given, but one argument that had been employed by President Clinton and other proponents of NAFTA was that the trade agreement would add an estimated 200,000 jobs to the American economy by 1995 alone. This projection came from an economic math model developed by Gary Hufbauer and Jeffrey Schott of the Institute for International Economics [2]. The Hufbauer-Schott model is based on a number of assumptions, the key one being that the United States would run trade surpluses with Mexico in the foreseeable future. On the other hand, the Prestowitz model, after Clyde V. Prestowitz, Jr. of the

Economic Strategy Institute, predicted that NAFTA would eliminate 420,000 jobs over the 1990s, based on the key assumption that Mexico runs trade surpluses with the United States.

After heated debate the North American Free Trade Agreement was passed by Congress in November 1993 and took effect on January 1, 1994.

Reality has not been kind to the projections of the Hufbauer-Schott model. The U.S. trade balance with Mexico went from a surplus of $1.7 billion in 1993 to a $16.2 billion deficit in 1996; the trade deficit with Canada went from $10.7 billion in 1993 to $22.8 billion in 1996. By using a method of calculation developed by the Commerce Department whereby each $1 billion in exports is equated to 14,000 jobs, 420,000 jobs have been lost as a result of the change in the U.S. trade balances with Mexico and Canada. As of March 1997, the Labor Department had certified 107,632 U.S. workers as having lost their jobs to NAFTA and as being eligible for its NAFTA Trade Adjustment Assistance program. This figure provides another dimension to job loss due to NAFTA, but it cannot be viewed as all-inclusive because NAFTA-TAA is only available to some workers in some industries and workers file for assistance under other trade unemployment assistance programs as well. Reality's results make clear that NINO had spoken and that the Hufbauer-Schott model was considerably off reality's mark.

Are the differences between the Hufbauer-Schott and Prestowitz models "significant" in terms of jobs gained or lost because of NAFTA? The answer depends on one's vantage point. From the point of view of Labor and workers and their families who stand to be uprooted, the realism of these models is highly significant. The view held by some economists was that in terms of the then nearly $8 trillion U.S. economy, NAFTA was economically trivial and that the treaty should be supported because it is an important step in the direction of "free" trade. Many pro-NAFTA politicians favored NAFTA because of what they saw as its political benefits. Richard Feinberg, formerly an advisor to President Clinton, is quoted as remarking: "We bought ourselves an ally with NAFTA" [3]. Now the question is, which model gave best support to a pro-NAFTA political outlook? The answer, of course, is the Hufbauer-Schott model. Its favorable job creation projection made it the favorite of the pro-NAFTA political and "free" trade economic establishments. For them the realism of its assumptions was of no concern.

THE CONSUMER PRICE INDEX MUST BE REFORMED!

In 1995 center stage of government business was taken up by the problem of eliminating the federal budget deficit. As part of their plan to accomplish this, House Republicans recommended that the annual cost-of-living adjustment for Social Security and other benefits tied to increases in the consumer price index (CPI) (see "Coming to Grips with Inflation," Chapter 11) be reduced starting in 1999 [4]. Washington wisdom circulating at the time held that the CPI overstated inflation by as much as 1.5 percentage points and that a reduction in the CPI's value was not only justified but defensible.

The problem was to give legitimacy to the Washington wisdom. Since the Bureau of Labor Statistics (BLS) moved cautiously and, from the point of view of the Washington establishment, unreliably in this matter, in June 1995 the Senate Finance Committee appointed a five-member panel of economists, chaired by former President Bush economic advisor Michael Boskin, to study the CPI and make recommendations on revisions. All members of what came to be called the Boskin Commission had respectable credentials and some might be described as eminent, but all had previously given congressional testimony that the CPI exaggerated inflation. Economists who took a different view, such as former Commissioner of BLS Janet Norwood, were not invited to join the panel.

The Boskin Commission released its report in early December 1996, claiming that consumer inflation was being overstated by the CPI by about 1.1 percent a year, arguing that the index did not adequately reflect the improving quality of goods, did not take into account new products quickly enough, did not properly reflect consumers' tendency to purchase cheaper alternatives when the price of goods rose, and did not properly take into account consumer shift toward discount stores [5, 6].

A number of questions arise. What does it mean to say that the CPI overstates the reality of inflation by about 1.1 percent a year? Many interpret this to mean that there is an ideal standard for measuring the reality of inflation which is known by the Boskin Commission and that in comparing the BLS's CPI against this ideal standard, the BLS's CPI overstated inflation by about 1.1 percent a year. This is utter nonsense; there is no ideal standard. The BLS's CPI is a valid conclusion of a math model based on data, accepted procedures, and assumptions made by the agency's economists. The Boskin Commission's proposed 1.1 percent per annum adjustment is based on the same data, same procedures, but with somewhat different assumptions. In effect, what the Boskin Commission was saying was that if you employ our assumptions rather than your's, then you have to make a 1.1 percent per annum downward adjustment in your CPI value. If your CPI increased by 3.3 percent over 1995, then 2.2 percent would be a more accurate description of the reality of inflation over that year, based on our assumptions.

Did the Boskin Commission consider the possibility that the BLS's CPI understates inflation? No. For discussion of this situation see Refs. 7 to 10.

Is there good reason to prefer the Boskin Commission's assumptions over those made by the BLS? If politics were put aside, it becomes "experts" vs. "experts." At a panel session at the annual meeting of the American Economic Association held in New Orleans in January 1997 Boskin and his four commission colleagues presented their views followed by BLS Commissioner Katherine Abraham who stated that she "agreed with some of the Boskin Commission's recommendations, including that the CPI should be as close to a measure of the cost of living as possible." She added, however, that her agency "would not and should not produce a CPI based partly on subjective judgments" [11]. Abraham later further elaborated to the Senate Finance Committee: "If we get into the business of making judgments about things that are not measurable—guessing, even if it's . . . a best guess—we really, I think, would be undermining the credibility of all of the data we produce" [11].

The planting of an idea which developed into a BOLÉRO drum roll to reduce the CPI was probably done by the Congressional Budget Office when it asserted in late 1994 that the CPI exaggerated inflation by an amount between 0.2 and 0.8 percentage points a year. Federal Reserve Board head Alan Greenspan expressed the view that the CPI exaggerated inflation by an amount between 0.5 and 1.5 percentage points a year at a joint meeting of the House and Senate Budget Committees in January 1995. Greenspan also noted that correcting these estimates could save the government $150 billion over five years and suggested the possibility that Congress pass a law that would lower the CPI by a percentage point or half a percentage point for determining benefits tied to the CPI. It's a short hop from this plateau to the establishment of the Boskin Commission and what became the "official" view that the CPI overstated inflation by about 1.1 percentage points a year.

The Boskin Commission's report set BOLÉRO into motion in the form of a rash of calls to "fix" the way inflation is measured. Testifying before the Senate Finance Committee on January 30, 1997, Alan Greenspan recommended that an independent commission be established to set cost-of-living adjustments for federal receipts and outlays each year.

Economist Martin Feldstein, who had been President Ronald Reagan's top economic advisor, suggested that Greenspan's proposed committee should recommend an "appropriate" inflation adjustment factor through informed judgment, apart from any adjustment made to the CPI by the Bureau of Labor Statistics through its normal work. Senators William Roth and Daniel Patrick Moynihan introduced a sense of the Senate resolution that urged an accurate cost-of-living index.

With momentum at a peak to push through a CPI fix, it all came apart. President Bill Clinton, faced by strong opposition within his own party and constituencies like Labor and the elderly, decided not to pursue a CPI fix outside of the highly professional, nonpolitical machinery of the Bureau of Labor Statistics. Republican enthusiasm for a fix waned with the discovery of a two-year old White House memorandum on how Democrats could use the issue against Republicans.

Within two years talk of engineering a CPI fix to help eliminate the budget deficit had turned to arguments over what to do with the projected budget surplus.

SOCIAL SECURITY IS COLLAPSING

According to a poll conducted by Peter Hart Research in February 1998, 72 percent of those in the age group 18 to 34 expect Social Security to pay much reduced benefits or no benefits at all when they are ready to retire. For working Americans in general the figure was 60 percent [12]. These figures are a tribute to the success of BOLÉRO and the art of hyping a solvable problem concerning Social Security's financing into what appears to be a time bomb that will wreck devestation on America's elderly when it explodes around 2030—"If we don't reform Social Security it could be running an annual deficit of $700 billion to as much as $1.3 trillion annually," stated a full-page ad which appeared in *The New York Times* in the fall of 1996.

In what may be termed a five-year cycle for forecasting gloom and doom on Social Security, President George W. Bush's hand-picked commission on Social Security warned that 2016 is Social Security's crisis date—the year the program's obligations exceed its payroll tax revenues, so that it begins to run a deficit instead of a surplus. This time bomb must be defused, became the war cry, which opened the door to a number of proposals to save Social Security and the problem of saving Social Security from its would-be saviors.

What justifies the time-bomb image of Social Security? Mathematics does, it is argued, and sometimes "mathematics" is prefaced by simple or straightforward, and repeated in a tone which suggests that only a fool or simpleton would be willing to quarrel with mathematics, particularly when it's simple or straightforward.

At the risk of being branded simpletons, we should be ready to quarrel with mathematics, not the technical part which comprises the deductive proof, but in the sense of taking a close look at the assumptions which underlie the math deduction. After the close look, we may or may not want to quarrel with the assumptions, but we should always keep in mind the hazards of long-range forecasts. (See "The Hazards of Long-Range Forecasts," Chapter 11.) The fact is that the time-bomb projection made by the Advisory Council on Social Security (SSA) rests on what Aaron Bernstein of *Business Week* termed "gloomy assumptions about population and labor force growth that aren't shared by other government forecasters." "They're also difficult to square with historical trends," Bernstein noted [13]. Other key factors are the growth rates of gross domestic product (GDP) and productivity.

The average-annual growth rate of inflation-adjusted GDP and productivity over the next 75 years are assumed by SSA to be 1.4 and 1.3 percent, respectively, which are considerably gloomier than the 3.0 and 2.1 percent growth rates that we've had over the past 75 years. As Bernstein notes, [14], "if you buy the notion that we're in for 75 years of 1.4 percent growth, you have a lot more to worry about than just Social Security." Bernstein also points out that SSA has an alternate model in which

GDP growth is assumed to average 2.14 percent a year over the next 75 years and that in this model Social Security is likely to show a slight surplus through 2072.

What becomes more and more clear, as a *Business Week* editorial pointed out [15]: "A modest long-term problem is being hyped into an impending catastrophe. And battle lines drawn in Washington over Social Security have less to do with pragmatic solutions than with liberal-conservative wars dating back to the Roosevelt Administration." For further discussion see, for example, Refs. 16 to 30.

13

Lessons from Chapters 8 to 12

PERSPECTIVE ON MATH PROOF

We return to math views 4–7 by posing for them the same questions we posed for math views 1–3 considered in Chapter 7. Do these views express math insights, math myths, or a bit of both attractively packaged as mathematical Trojan horses?

4. *Mathematical proof is the most reliable means for objectively establishing truth.*

This is an attractively packaged math myth. It appeals to our yearning for something solid, indisputable, in a world beset by relativism and conditionals. As Bottom-Line Bob put it to Reflective Ramunė: "What I like most about math is its precision. You don't get ten sides to a story. You get one answer and that's that; no baloney." (See "'Precise' Mathematical Reasoning According to Bottom-Line Bob," Chapter 10.) As Bob learned, much to his discomfort, mathematical life is not all that simple. Ramunė's reply to Bob must have come as quite a surprise to him: "Mathematics is precise in the sense that it gives us valid conclusions with respect to the assumptions made. Whether the assumptions made are realistic or not is another matter. . . ."

5. *The precision of mathematical methods (that is, the techniques used to construct a "math proof") guarantees unassailable conclusions that serve as pillars of stability and strength in a world besieged by foggy thinking, prejudice, and rampant special interests.*

Jules Warner's Uncle George believes that unassailable conclusion means true statement, as he made clear in his pronouncement about the precision of math proof in geometry. (See "'Precise' Mathematical Reasoning According to Uncle George," Chapter 8.) As Jules would have put it had he not given first priority to getting a drink: No. The precision of math methods guarantees unassailable conclusions in the sense of validity, not truth. The rest of the verbal color is just that, verbal red herrings.

6. *Mathematically derived conclusions are indisputable because they are based on deductive logic, which is untainted by bias and ideology.*

Whether we use the term *indisputable* or *unassailable*, as in statement 5, it comes down to the same point. Math conclusions are indisputable only in the sense of validity.

7. *Mathematical methods have the advantage of being certain. No qualified person can resist the truth of a mathematical conclusion properly communicated. The job of communication may be difficult if the solution is complex, but when the communication is competent, agreement is inevitable.*

This version of the nature of math methods is even more loaded with red herrings than its companions, but it comes down to the same issue: truth in the real world sense vs. validity in the deductive logic sense.

Mathematical methods have the advantage of being certain, it says. But certain in what sense? Although certain in the sense of real-world truth might be the answer we long to believe, certain, once again, means certain in the sense of validity. The phrase truth of a mathematical conclusion refers, of course, to its validity.

Many writers use the phrase truth of a mathematical conclusion for its truth in the world of mathematics, meaning validity, rather than real-world truth. One may lament that such usage promotes confusion and should be avoided, but the fact is it's become common practice. This means that the burden of distinguishing the sense of truth intended is shifted to us, the readers.

BUILDING A DEFENSE AGAINST QUESTIONABLE MATH MODELS

Application of the following four W's to mathematical conclusions claimed to be based on studies should put us in a better position to detect questionable claims and math models.

W1. Watch the assumptions.

Often our first task will be to dig out the assumptions. In the spirit of out of sight, out of mind, they are usually hidden. If we succeed in finding them, we can give attention to their realism. In carrying out this task, we should not overlook what was referred to in "Additional Strings Attached," Chapter 8, as postulate P: Other factors which may play a role in the situation being studied are not being considered. Is postulate P realistic, we should ask ourselves, or are there significant factors which are unrealistically being treated as negligible.

The process of examining the realism of the assumptions is often referred to as validating the assumptions. In this context the term *validate* has no connection with *valid* as used in a deductive logical or mathematical sense. The opportunity for confusion to raise its ugly head is great, which, alas, puts the burden of keeping in proper perspective the sense in which *valid* and *validate* are being used squarely on us.

If the assumptions are pinpointed and found not to be realistic, or if they cannot be identified, it would not be wise for us to put high hopes on the realism of the conclusions being held up as following from them.

W2. Watch the wording.

The terms *projection*, *forecast*, *estimate* are often used instead of *conclusion*. The word *estimate* is particularly elusive because we tend to make allowances for the inexactness communicated by *estimate* that we might not be inclined to make for the more precise-sounding *projection* and *forecast*. These terms are often used synonymously. There are good estimates, which are in some sense close to reality's mark, and poor estimates, which are distant from reality's mark. This brings us back to W1.

W3. Watch out for long-range forecasts (projections, estimates).

As discussed in "The Hazards of Long-Range Forecasts," Chapter 11, long-range forecasts are quite often considerably off reality's mark. This makes careful examination of the assumptions underlying such forecasts and the realism of postulate P especially urgent priorities.

W4. Watch out for numerical quicksand masquerading as bedrock in connection with math modeling.

It is helpful to view the cryptic-sounding W4 against a specific situation. The setting that prompts W4 is the campaign to force a lowering of the consumer price index value (see "The Consumer Price Index Must Be Reformed!" Chapter 12) which unfolded in the middle 1990s.

In 1995 a commission of economists appointed by Congress (the Boskin Commission) issued a report stating that the CPI value overstated inflation by about 1.1 percentage points annually.

This view, so seemingly solid because of the sense of mathematical bedrock it communicates, is more appropriately viewed as quicksand because it presupposes there is an ideal standard against which the CPI value was measured and found wanting to the extent of a 1.1 percent annual overstatement. There is no ideal standard. As we have noted, the CPI value is the valid conclusion of a CPI math model based on assumptions made by economists and statisticians of the Bureau of Labor Statistics. The Boskin Commission was in effect saying that if we were formulating the assumptions of the CPI model, the CPI value that would emerge as a valid consequence would be about 1.1 percentage points less. Viewed from this perspective, it becomes clear that the issue centers on the question of what assumptions should be taken as the basis for a CPI model. This, of course, is a matter of judgment.

The argument that the CPI should be revised downward by 1.1 percentage points because the Boskin Commission has "superior" judgment is untenable. But if the 1.1 percent figure is anointed as the most authoritative word on the subject, the case for adjusting the CPI downward by this amount through congressional mandate assumes greater weight—in fact, a necessity driven by logic is suggested. This approach to reducing the CPI was attempted but collapsed when it became clear that those leading the charge would be committing political suicide.

CONSIDER, FOR EXAMPLE, . . .

Yankee Stadium in Manhattan?

In the late 1990s New York Mayor Rudolph Giuliani and his administration were giving serious thought to Manhattan as the future home of the New York Yankees. A new stadium on Manhattan's west side would have an economic impact worth at least $1 billion a year, a study prepared by the city's planning and economic—development departments estimated [1]. One billion dollars is an enticing figure and Mayor Giuliani was quick to seize on it to promote the idea of a Manhattan Yankee Stadium: "This spending will result in, bare minimum, at least $1 billion more for our economy and thousands and thousands and thousands of jobs. We need those jobs" [1].

For the doubting Thomases, Janes, and the rest of us as well, this should prompt the question: What basis does the $1 billion figure have in reality? The planning directors who authored the report declined to give details about the methodology used and assumptions employed. This by itself, of course, should signal a red alert.

Based on a study it conducted, KPMG Peat Marwick concluded that a new baseball stadium for the Yankees on the west side of Manhattan would generate about $100 million for New Yorkers [2]. The same questions about methodology and assumptions pertain.

Considering the tendency of businesses and local governments to put the most positive spin on figures they want to convince us of and the difficulty in obtaining reliable data and setting up realistic assumptions, New York City's former budget director Carol O'Cleireacain spoke for many of us when she commented: "As an economist, I wouldn't believe any study I saw" [2].

We've Got $0.75 Trillion, or Is It $0.84 Trillion, or Is It $1.86 Trillion, or Is It . . . ?

In February 2000 President Bill Clinton projected a surplus of $0.75 trillion for the non-Social Security part of the budget over the 10-year period 2001–2010, assuming that spending grows at the rate of inflation. The Congressional Budget Office projected a surplus of $0.84 trillion, with respect to the same key assumption; a surplus of $1.86 trillion with respect to the key assumption that spending is frozen at the level for the 2000 fiscal year; and a surplus of $1.92 trillion with respect to the assumption that spending follows the 1997 budget caps through 2002 and then grows at the rate of inflation [3].

Which is correct? All and probably none. All are correct in the sense of validity. Which is close enough to the actual surplus obtained by adding up the figures obtained for the 10-year period 2001–2010 to be labeled true? We will not be able to definitively answer this question until 2010, but there is a strong likelihood that the answer is none. If you are inclined to wager that one of these projections will be "reasonably close" to reality's mark (suitably defined), then before doing so I would urge you to review "The Hazards of Long-Range Forecasts," Chapter 11, on the hazards of long-range forecasts and insist on astronomical wagering odds to make it a fair bet.

Cost of Medicare Coverage for Prescription Drugs?

When President Clinton announced his drug coverage proposal in June 1999, he said it would cost $118 billion over 10 years. The Congressional Budget Office estimated the cost at $168 billion, 42 percent higher than President Clinton's estimate [4].

Which, if either, is close to reality's mark? Again it comes down to the realism of the assumptions being made with the wild card being provided by the hazards of long-range forecasts.

BUILDING A DEFENSE AGAINST HYPERBOLE AND HYPE

The problem faced by those seeking to promote a point of view or public policy position is basically the same as that faced by those seeking to sell a brand of soda, automobile or what have you to the public. The techniques employed are, with suitable refinements, much the same. The best defense against hype in public position selling is to always keep in mind that the key verb is to sell, whatever the product being pushed, and place the defense mechanisms you have developed against being sold a bill of goods on red alert.

Endorsements

One approach to capturing the sneaker market with your brand is to gain the endorsement of a popular sport's figure to speak on its behalf. For selling a public policy position a well-known, respected public figure would be the appropriate choice. The idea is the same: If basketball star, olympic champion, Doctor, President, Senator, etc. Jones says it's "good," then you should buy into it.

A variation on this approach is to send a signal through numbers of endorsements, the idea being that if so many well-thought-of people think it's good, then it must be so. Consider, for example, the full-page ad floated in support of NAFTA: "WHY 300 ECONOMISTS, 136 NEWSPAPER EDITORS, AND 6 PRESIDENTS SUPPORT NAFTA: NAFTA WORKS FOR EVERYONE" [5]. Five claims were made in support of

NAFTA and the thrust of the ad was that we should buy into NAFTA because so many economists, newspaper editors, and former presidents have done so. It sounds impressive, but whenever I see such an ad I'm prompted to think back to a time when a similar ad endorsing the idea that Earth is flat could have been floated.

Image Coloring

There are a number of words that have come to evoke a positive or negative emotional response from us. Words such as *free, democratic, bipartisan, reform,* and *expert* elicit positive images, whereas *corrupt* and *greed* elicit negative ones. To minimize the chance of being unduely influenced by hype we should be careful of the coloring added by the appearance of such words. For example, NAFTA stands for North American Free Trade Agreement. *Free* in NAFTA does not mean *free* in the sense that many of us associate with the word. NAFTA spells out details for carefully managed trade, but the more accurate NAMTA would be a harder sell than the less accurate but more appealing NAFTA.

The Boskin Commission was a five-member panel of economists chaired by former President Bush economic advisor Michael Boskin that was appointed by the Senate Finance Committee to study the consumer price index and make recommendations on revisions that would bring the CPI into "closer" contact with the reality of inflation. (See "The Consumer Price Index Must Be Reformed!" Chapter 12.) Generally, editorials on revising the CPI and commentaries on the subject described the Boskin Commission as "blue-ribbon," "distinguished," and "bipartisan." Such characterizations promoted the view that the Boskin Commission's analysis and conclusions were wisdom being handed down from on high, which promoted their uncritical acceptance by many.

Reality, however, was more complex. All members of the Boskin Commission had expressed the view that the CPI overstated the reality of inflation prior to their appointment to the commission, so that their conclusion came as no surprise. Only the amount by which the CPI overstated inflation remained to be specified and it was fixed as being about 1.1 percent per annum. No economists who had doubts about the CPI overstating inflation were appointed to the commission. The fact that the Senate Finance Committee which appointed the commission was bipartisan was not relevant to the commission's selection since Republican and Democratic members of the committee were in agreement that the CPI overstated inflation. The fix was in prior to any shuffling of the papers taking place. The commission conducted no new research and its findings were not subject to professional review prior to being released. For further discussion, see Refs. 6 and 7.

Some discussions of the Social Security issue have taken to referring to the elderly as the greedy elderly, which is a device intended to make more palatable the idea of cutting benefits for the elderly.

Part Three

THE PRICE TAG

INTRODUCTION

Unrealistic conclusions and numbers breed poor decision making which may result in substantial waste of money, squandered resources, diminished quality of life, and missed opportunities.

Chapters 14 and 15 consider the price tag ramifications of a variety of situations. Also considered in Chapter 15 is the false economy achieved from penny-pinching on data collection.

14

The Numbers You Don't Know Can Hurt You

SIDE EFFECTS? WHAT SIDE EFFECTS?

Halcion, manufactured by the Upjohn Company and introduced in the United States in 1983, is one of the world's best-known sleeping pills. Its main advantage over competing products, Upjohn claimed, is in encouraging nighttime sleep without daytime drowsiness.

How safe is Halcion? It received Food and Drug Administration (FDA) approval, and its manufacturer claimed that it is just as safe as other drugs of its kind. Dissenters argued that Halcion is more likely to cause symptoms such as amnesia, paranoia, and depression and that Upjohn engaged in data manipulation to conceal its side effects. This view emerged from a law suit filed by Ilo Grundberg, who killed her mother the day before her mother's eighty-third birthday and placed a birthday card in her hand. Mrs. Grundberg claimed that Halcion had made her psychotic, and charges against her were eventually dismissed. (The author William Styron describes his experience with Halcion in his book *Darkness Visible: A Memoir of Madness* [1].) Upjohn settled the lawsuit with Ilo Grundberg before it was to go to trial in August 1991, but in preparation for the suit it had to make available a good deal of data about Halcion to her attorneys.

Dr. Ian Oswald, who was head of the department of psychiatry at the University of Edinburgh and spent 30 years doing research on sleep, was obtained as an expert witness. Dr. Oswald spent two years going over Upjohn's data and concluded that Upjohn had known about the extent of the drug's adverse effects for 20 years and concealed these data. He concluded that "the whole thing had been one long fraud" [2]. Dr. Graham Dukes, former medical director of the Dutch drug regulatory agency, who examined some of Upjohn's data, believed that the data on Halcion had been organized in a way to minimize the drug's adverse effects and that this could not have occurred accidentally.

In reaction to the criticisms voiced, Britain, the Netherlands, and Belgium were led to remove the sleeping pill from the market. A report issued in April 1994 by FDA investigators stated that the Upjohn Company had engaged in ongoing misconduct with Halcion.

TELL MORE, OR HUSH HUSH?

Derivatives

Derivatives are contracts that allow investors to bet on the future financial performance of things that change in value over time, such as interest rates, stocks, currency rates, grain, pork, and any commodity whose value can be expected to change in the near future. Investors buy and sell these contracts, whose value changes depending on changes in the value of the underlying product. If you buy a derivative based on the expectation that the dollar will be worth 1.25 euros next month, when its value is 1 euro now, and the dollar falls to 0.8 euros, you stand to lose your shirt; if the dollar goes to 1.25 euros, you stand to clean up. A few years back Orange County, California, lost more than $1 billion by investing in derivatives, making very clear that are they not for the fainthearted.

Should companies be required to show derivatives transactions on their balance sheets? The Financial Accounting Standards Board (FASB) and the Securities and Exchange Commission (SEC) believe that the answer is yes: "Financial markets demand transparency. Having billions of dollars of derivatives transactions unaccounted for creates an intolerable risk for investors" [3]. The FASB and SEC position was opposed by financial institutions which cited "concerns related to the potential impact on the capital markets, the weakening of companies' ability to manage risk, and the adverse control implications of implementing costly and complex new rules" [4].

In May 1998 the FASB gave final approval to a rule that would require companies to record the market value of derivatives on their balance sheets and to include the gains or losses on these contracts in income. Edmund L. Jenkins, chairman of the accounting board, noted: "Unfortunately, there are some who have been trying to force the FASB not to issue this standard, preventing valuable information from reaching investors. We must avoid placing the interests of any particular group over the consumers' interests" [4].

Tobacco

Internal documents from the Brown & Williamson Tobacco Corporation going back to the early 1960s reveal a very different picture from the public posture that we have become accustomed to hearing. In 1963, when Dr. Luther Terry was preparing the first Surgeon General's report warning against cigarette smoking as a health hazard, an internal debate raged over what research data in the company's possession should be disclosed to the Surgeon General and the public position that should be adopted.

In July 1963 Addison Yeaman, then general counsel for the company, suggested that the company "accept its responsibility," inform the Surgeon General about what it knew about the hazards of cigarette smoking, and pursue research to develop safer cigarettes.

This counsel was not followed and the tobacco industry pursued a public relations strategy emphasizing the smoker's freedom of choice and the glamour and sophistication of smoking, while denying that health hazards had been proved.

THE "VALUE" OF A LIFE?

On February 4, 1993 a state court jury in Atlanta held General Motors responsible for the death of 17-year-old Shannon Moseley in an accident in which his 1985 GMC Sierra pickup truck burst into flames when it was struck from the side. The jury determined that GM knew the truck's design was flawed, but kept the danger from the public. It awarded $4.2 million in actual damages and $101.2 million in punitive damages to the parents of the boy who was killed. Key testimony was provided by Ronald Elwell, a former GM safety engineer, who testified that the company withheld crash test data from him until 1983 and that he continued to defend GM until 1989 out of fear of losing his job.

Citing technical points, a Georgia appellate court overturned the verdict in June 1994, but held that the evidence submitted showed that GM was aware that the gas tanks posed a hazard but did nothing to make them safer.

In October 1994 Transportation Secretary Federico Peña announced an initial finding that GM's pickup trucks posed a fire hazard when struck from the side and further stated that GM management at the time appears to have made a decision favoring sales over safety. In December 1994 Mr. Peña announced a settlement with GM that would end the federal investigation in exchange for GM's promise to contribute $51.3 million to auto safety programs.

DANGEROUS NUMBERS

On March 20, 1996 a public service notice appeared in *The New York Times* which read, in part [5]:

> My name is Antonio Benedi. I took Tylenol as directed and lost my liver. I came within hours of death and was saved only by an emergency liver transplant. Later, I won a lawsuit against the McNeil company.

Startling, to say the least. After all, Tylenol is the most popular over-the-counter pain relief medication in the country, with 8 billion odd tablets taken annually by Americans. Any medication taken in excess will cause side effects, but Mr. Benedi, who admitted to being a moderate consumer of alcohol, followed directions, and lost his liver. The question of course is, what's the story?

The story is that acetaminophen, the active ingredient of Tylenol, becomes toxic at doses close to that used in treatment. Overdoses of medications containing acetaminophen were the leading cause of hospitalization for acute liver failure at Parkland Memorial hospital in Dallas from 1992 to 1995, a study found [6].

More generally, in 1994 over 100,000 people died in American hospitals from adverse reactions to prescription medication, it has been estimated; more than 2 million suffered serious side effects. This makes adverse reactions to prescription drugs the sixth leading cause of death in the United States in 1994, below heart disease and cancer, but above diabetes and HIV/AIDS [7,8].

SHIELD WITH A HOLE

One day in November 1992 Diane Ayres took a single pill of a new antibiotic, Omniflox, which had been prescribed by her doctor for a minor urinary tract infection. Six hours later, delirious, she was brought to the emergency room. Diane Ayres had had a severe reaction to a legal medication that had Food and Drug Administration (FDA) approval [9]. How could such a thing happen?

The fact is that FDA-approved drugs enter the market with their long-term safety unknown. [The most extreme example of this to date is the drug DES (diethylstilbestrol), prescribed for threatened miscarriage, which reached beyond the pregnant women who took it to wreck havoc with the lives of their children.] In a sense the general public is a collection of guinea pigs for the further testing of drugs. The General Accounting Office estimated that 51 percent of FDA-approved drugs have major adverse effects that are discovered after they had been released to the public [10].

Be that as it may, the suggested follow-up question is: What is the safety net for recording and reporting adverse reactions to FDA-approved drugs that have entered the market? The news is disturbing. "When Dr. Brian L. Storm gives talks to medical groups, he shows a slide of the Food and Drug Administration's form for reporting adverse reactions to drugs. How many of you have ever seen that? he asks the doctors in the audience. On average, fewer than a third raise their hands." "The real issue, probably more important than the reporting system, is for people to be looking, to be on guard, to be looking for possible associations of drugs with adverse reactions," noted Dr. David A. Kessler, former commissioner of the FDA [11].

15

Consequences of Math Fiction

A FAULTY, NO-FAULT DIVORCE STATISTIC

In her 1985 book, *The Divorce Revolution,* Stanford University sociologist Lenore Weitzman stated that an economic consequence of no-fault divorce is that women's standard of living drops an average of 73 percent in the first year after divorce, whereas men's rises 42 percent. Although sociologists challenged these figures almost immediately, they were the figures that stuck. They were cited in more than 100 national magazines and newspapers, including *The New York Times* and *Cosmo*, at congressional hearings, in more than 250 law review articles, in at least 24 state appellate and Supreme Court cases, and by the Supreme Court itself [1]. They were the figures of choice of critics of no-fault divorces seeking to overturn or block no-fault divorce legislation. The fact that they were inconsistent with other findings and had not been verified mattered little, if at all. A hot statistic was too good not to be true.

Weitzman's figures proved to be false. In 1993 sociologist Richard Peterson obtained access to Weitzman's data, reanalyzed them using the same methods she had employed, and obtained figures of 27 percent for the decline in the standard of living of women and 10 percent for the rise in standard of living of men. These results were in line with those that had been obtained by other researchers. Weitzman believes that her erroneous results were due to problems with her computer files.

Peterson's observations [2] that "the discussion of no-fault divorce and other legal reforms has been seriously distorted by . . . inaccurately large estimates. To be effective, these reforms must be based on reliable data," should be taken seriously for their general applicability.

DRUGGED DRUG NUMBERS

In the early 1990s managed care hit the pharmaceutical industry with an impact that brought about a quiet but troubling revolution in the testing of drugs with the potential to fight disease. With managed care in orbit, drug companies found themselves under enormous pressure to hold down prices. Until this point they had felt free to set drug prices almost at will, and it was from this source that they derived the bulk of their profits rather than scrambling to develop new drugs. But now the pressure to develop new drugs had become intense.

Time is money, the old adage has it, and the clinical trial part of drug development conducted by research facilities and hospitals was time consuming, to say the least. Drug companies more and more began to recruit private-practice doctors who were willing to cull their patient listings for research subjects. In 1997 about 70 percent of the 16,000 new trials were initiated by private-practice doctors rather than university research facilities or hospitals.

At first sight this might seem like a win-win situation, good for all concerned. But like many things that seem too good to be true, this one was. A 10-month investigation carried out by *The New York Times* pinpointed the following concerns [3]:

- Drug companies offer large payments to doctors and medical personnel to encourage them to recruit patients for trials to be conducted.
- A premium is placed on speed and meeting quotas.
- The system is fraught with conflicts of interest. The only thing separating a trusting patient from a study that might be inappropriate or potentially harmful is the judgment of the doctor, who is being subjected to powerful temptations.

To correctly distinguish between drug reactions and disease symptoms a doctor's specialized training and experience are invaluable. But clinical trials are often conducted by doctors whose specialty is far removed from the condition troubling the patients under study—psychiatrists carrying out Pap smears, asthma specialists, administering psychiatric drugs, and the like. As to experience, the *New York Times* study showed that during the 1990s, 70 percent of the doctors conducting trials involving people had previously been involved with three or fewer drug studies.

And then, as the Fiddes case makes clear ("The Drug Testing Balloon That Popped," Chapter 1), the potential for fraud on a mass scale is opened up. The popping of the Fiddes balloon left in its wake compromised drug-study results for almost every pharmaceutical house in the business.

The basic underlying issue centers on the integrity of the data—which, of course, translates to integrity of the numbers generated from the studies. At the end of the road we all pay part of the tab which may come in the form of monetary cost (in many ways the simplest of the costs), quality of life for ourselves, family, friends, members of our community, and life itself.

ADVERTISING, TV OPTIONS, AND QUESTIONABLE NUMBERS

The life span of a television program is determined by the public's reaction to it, which is measured by TV ratings. These ratings, produced by the Nielsen Company, estimate the audience in terms of the percentage of those sets in use which are turned to each channel, called a *share*, or in terms of the percentage of the total possible audience, sets on or off, called a *rating*. Shares and ratings are further broken down according to the sex and age of viewers so that advertisers can better focus their advertising campaigns. These numbers determine the buying and selling of billions of dollars of television air time. They mean life or death to television programs. The half-hour comedy *Good & Evil*, which had promising ingredients in terms of writing, acting, and production talent, had a short life after its premiere in the fall of 1991 because of low initial ratings. In March 1992, NBC announced that they were dropping two successful shows, *Matlock* and *In the Heat of the Night*, because the demographic numbers favored older viewers while the network wished to build around a more youthful audience.

Since 1986 the data that underlie the ratings have been collected by a device called a people-meter. The remote control part of a people-meter rests on top of the television set and when the set is turned on the meter prompts viewers to enter their identification number. Information is provided on what channels are being beamed into the household and who is watching them. Nielsen puts its people-meter into 4000 households selected at random—that is, without bias—from the approximately 93 million homes in America with television.

The people-meter data-gathering system produced lower ratings for the networks than had been expected, and a serious question arose as to

whether this was because of the increased or decreased accuracy of this system over the method it replaced. The networks commissioned a study of the Nielsen methodology and two years later this Committee on Nationwide Television Audience Measurement (CONTAM) issued a nine-volume report that was highly critical of the Nielsen system. The report found evidence of button fatigue—that over time people did not push the buttons that would ensure data accuracy as they did in the beginning. CONTAM was highly critical of Nielsen's sampling procedures for obtaining the 4000 households that make up their sample; random sampling was envisioned in the methodology, but the actual sampling deviated significantly from this requirement. From this came ratings that were highly suspect.

Nielsen overcame the statistical sampling problem, but it was still plagued by the problem of getting "honest" data from viewers in the sample selected. Its people-meter system for eliciting viewing data was described as too mechanical and as not being user friendly. The problem of obtaining accurate viewer data remained. Matters came to a head in March 1997 with results of what is termed the February sweeps, an intense ratings period that determined television's winners and losers in terms of how $46 billion in advertising money was to be allocated.

According to the Nielsen ratings, the average number of American households watching prime-time television fell by over 1 million in February 1997 compared to February 1996. This was the fourth decline in the preceding five years. The networks did not find Nielsen's numbers credible. As Don Ohlmeyer of NBC put it, "I don't trust their numbers at all. They're trying to measure twenty-first-century technology with an abacus" [4]. Nielsen's response was that the networks were engaging in the time-dishonored practice of blame the messenger. It is safe to assume, however, that they undertook to improve their data collection system.

In December 1998, Nielsen reported that the 18- to 34-year-old television audience had fallen by 6 percent in the fall compared with its previous fall level. Nick Schiavone, senior vice president of research for NBC, described Nielsen's measurements as seriously flawed and estimated that NBC had lost $66 million dollars in advertising revenue during the fall season because of Nielsen's audience data [5]. Barry Cook, senior vice president of research for Nielsen, took issue with the network's view of flawed data, pointing out a long-term trend line that showed a dropping off of young adult viewers that went back to 1991.

Irrespective of who is closer to reality's mark in this heated exchange, the lesson that poor-quality data or poor-quality reading of the data can cost big bucks is reaffirmed.

BALANCE THE BUDGET AND PAY OFF THE NATIONAL DEBT

The mantra "balance the budget" was taken up by the Republican leadership in the 1990s, and serious pressure was generated to enshrine this condition as a constitutional amendment. Vice President Gore added "pay off the national debt" as a war cry for his presidential campaign of 2000. Are the conditions revenue and spending in balance and zero (or near zero) national debt the "right" conditions, conditions that should be given unequivocal first priority to ensure prolonged prosperity? There are good reasons for negative answers to these questions (see "A Precursor of Economic Collapse?," Chapter 5, and "Balance the Budget," Chapter 12), but valuable perspective is also added to the issue by looking at history.

Professor Frederick C. Thayer notes that there were six major periods in American history that began with budget balancing and debt reduction and ended with depression [6].

- 1817–1821: The national debt was reduced by 29 percent. The first acknowledged major depression began in 1819.
- 1823–1836: The national debt was reduced by 99.7 percent to $38,000, a virtual wipeout as Thayer put it. A major depression began in 1837.
- 1852–1857: The national debt was reduced by 59 percent. A major depression began in 1857.
- 1867–1873: The national debt was reduced by 27 percent. A major depression began in 1873.
- 1880–1893: The national debt was reduced by 57 percent. A major depression began in 1893.
- 1920–1930: The national debt was reduced by 36 percent. The sixth major depression, the Great Depression, as it came to be called, began in 1929.

These data by themselves do not "prove" that balancing the budget and reducing debt cause depressions, but they do make clear that these measures are not sufficient to ward off depressions and suggest that if carried out as a first priority, no matter what the circumstances, may serve to turn a recession into a depression. The Great Depression provides some noteworthy lessons. A recession had begun to take hold in the spring of 1929, followed by the Wall Street crash in October—a bad downturn that was severely aggravated by bad monetary and fiscal policies. In 1932, at the worst possible time, President Herbert Hoover raised taxes to help balance the budget and "restore confidence." This was not a mindless decision, but rather a part of the orthodoxy of the time. It should give us pause to reflect on the damage implementation of flawed orthodoxy can do.

RIPPLES

The consumer price index (CPI) has a wide reach, and if it required a theme song my choice would be the one that goes: The kneebone is connected to the thighbone is connected to the hipbone is connected to . . .

Fully Indexed Programs. These are programs for which automatic increases in benefits are triggered by increases in the CPI. Social Security, received by nearly 45 million beneficiaries, is the best known of these programs. Others include railroad retirement, with about 800,000 beneficiaries, supplemental social security income, with about 6.5 million recipients; veterans' compensation; and federal military and civilian employee pensions, paid to about 4 million retirees. The official poverty line rises each year in accord with the behavior of the CPI, which affects about 26 million recipients of food stamps, 25 million in subsidized child nutrition programs, and 5 million with federal student grants.

Taxes. To protect taxpayers from the effects of inflation, taxes are adjusted in a number of ways. This includes tax brackets, which determine tax rates on income; personal exemption and standard deduction levels; earned income tax credit; limit on itemized deductions; and pension contribution limits.

Economic Statistics. Real growth in the gross domestic product (GDP) and productivity (i.e. growth adjusted for inflation) depends on the CPI.

The smaller the increase in the CPI, the smaller will be the additional benefits paid to beneficiaries, the smaller will be the cost to the government, the greater will be the tax obligation of taxpayers and government tax revenue, and the larger will be the GDP and productivity figures. While there is no such thing as a "perfect" measure of inflation, a good deal rides on making the CPI model as realistic a model of inflation as possible.

PENNYWISE AND POUND FOOLISH

How is it to be done? What can we do to support the research and data collection needed to make the consumer price index model as realistic as

possible? The answer does not lie in setting up a Boskin-type commission to give legitimacy to the political mantra of the day—the 1995 mantra being that the CPI overstates inflation and must be reduced to help balance the budget. This approach is not only worthy of inclusion in Chapter 4 ("Fudge Factories"), but because of its subtle quality could justifiably be made the star of the show. The answer lies in providing the resources needed to improve the quality of economic indicators and data in general. The simple fact is that bad data often have very bad consequences. For discussion see, for example, Refs. 7 to 10.

The problem is that obtaining sufficient funding to do a proper job of number crunching has become an annual battle with Congress. Murray Weidenbaum, chief economic advisor to President Ronald Reagan, was prompted to observe [11]:

> The real costs resulting from providing inadequate funding for one of the most useful parts of the bureaucracy—the data gatherers—are truly awesome. An inaccurately high report of inflation can trigger an avoidable policy of monetary restraint followed by needless declines in capital formation, production, and employment. Bad information on productivity can generate investor decisions out of sync with real trends in the marketplace.
>
> Underreporting exports can produce news of "record" trade deficits followed by adoption of tougher protectionist policies, which could undermine the growth prospects for the U.S. as well as those nations it trades with.

Part Four

PREVENTIVE MEDICINE: EDUCATION

INTRODUCTION

How can we minimize our vulnerability to being math manipulated by special interests, or seduced by an attractive sounding program based on what seem to be compelling mathematical arguments? Education is the answer, you might say, and in theory I agree. Translating theory into practice, however, is not so simple a matter.

Chapter 16 looks at the difficulties that arise in translating theory into practice, options for education through self-study, and the role of computers.

16

Mathematics Education

MATH EDUCATION FOR WHAT?

"You can't argue with mathematics; whatever career you're going to pursue, you've got to know your math," we often hear. These "wisdoms" have become generally accepted in our culture because of the views that mathematics is the epitome of precision and that it is employed in activities that affect our lives in many ways. The question that occupies the spotlight is: How should mathematics be taught so that our children are given every advantage to pursue success in their chosen careers and ensure that the United States maintains supremacy in a modern technologically driven world? This focus spurred an ongoing debate on how best to teach math.

In 1989 the National Council of Teachers of Mathematics (NCTM) published standards for the teaching of mathematics which is often referred to as the new-new math to distinguish it from the new math movement of the 1960s and 1970s. In a preface to its new-new math standards the NCTM denounced a "long-standing" preoccupation with computation and other traditional skills." The NCTM standards promote the use of calculators; a focus on real-world problems; a hands-on, project-oriented approach; group work and discussion of math concepts; and the integration of algebra, geometry, and trigonometry.

Has the new-new math met with success? Many would argue that the answer is no [1–7]. In California a group of more than 200 scholars banded together to urge Secretary of Education Richard Riley to stop endorsing the new-new math programs [8]. A growing number of parents are banding together to insist that their children be taught basic skills and that traditional math courses be restored. A low point in a public official's ability to handle math basics was demonstrated by Stephen Byers, Britian's minister for educational standards. Asked to multiply 8 by 7 in a radio interview, he replied, "I think it's 54" [9]. This helped spur a back to basics movement in Britain. The NCTM as well has been prompted to recommend giving more attention to the fundamentals of computation rather than focus on concepts and reasoning [10].

MATH EDUCATION FOR PUBLIC AFFAIRS

Much about this state of affairs might be considered troubling. But how is this related to the kind of math education that would help us to cope with slippery math in public affairs? Alas, nothing and everything. Nothing, in that in the continuing debate about math education, *success* means success in terms of career preparation or keeping the nation technologically strong and competitive. Everything, in that the kind of math education that would equip us as citizens in a democracy to make choices presented to us in a cloud cover of slippery math isn't given a second thought. Math education that would equip us to handle such a basic task as multiplying 8 by 7 and the math that arises in the worlds of business, economics, finance, engineering, the sciences, and technology does not prepare us to handle the slippery dimension of math we encounter in public affairs.

In a curious way the more technical math we study, the greater the chance that math views 1–7 (discussed in the first sections of Chapters 7 and 13) will take hold, leaving us less prepared to handle the slippery dimension of mathematics.

One way to prepare ourselves for the slippery dimension of mathematics is through reading and self-study. The following books are useful for helping us become more aware and sensitive to number pollution and the slippery dimension of math modeling.

1. W. J. Adams with R. B. Adams, *Get a Grip on Your Math* (Dubuque, IA: Kendall/Hunt, 1996). Part One, "Get a Grip on Your Numbers" (chaps. 1–9), discusses sources of slippery numbers with the objective of helping the general reader become more aware and sensitive to their origins and nature. Part Two, "Get a Grip on Math Models" (chaps. 10–20), discusses the development of math-model portraits for a variety of phenomena, including astronomy, business and economics, radiocarbon dating, space, and the use of the math modeling process as a tool for inquiry. More generally, *Get a Grip on Your Math* examines such questions as: What can mathematics do for us? What are its limitations? In what sense is mathematics precise?
2. W. J. Adams with R. B. Adams, *Get a Firmer Grip on Your Math* (Dubuque, IA: Kendall/Hunt, 1996). This sequel to *Get a Grip on Your Math* is intended for readers who wish to dig deeper. It provides food-for-thought questions to help one obtain a more concrete understanding of slippery numbers, and a more in-depth discussion of ideas taken up in *Get a Grip on Your Math*.
3. J. Best, *Damned Lies and Statistics: Untangling Numbers from the Media, Politicians, and Activists* (University of California Press, 2000). Review comments have been positive. "*Damned Lies and Statistics* is a clearly written primer for the statistically impaired," noted Steve Weinberg in *The Christian Science Monitor*.
4. V. Cohn, *News & Numbers: A Guide to Reporting Statistical Claims and Controversies in Health and Other Fields*. (Ames: Iowa State University Press, 1989). "This guide is an attempt to explain the role, logic, and language of statistics, so we reporters can ask better questions about the many alleged facts of findings that rest, or should rest, on some credible numbers."
5. A. K. Dewdney, "*200% of Nothing: An Eye-Opening Tour through the Twists and Turns of Math and Innumeracy* (New York: John Wiley & Sons, 1993). "The first part of this book illustrates the most typical abuses of mathematics with examples drawn from the real world. The second part of the book looks through the other end of the telescope, so to speak. It examines the areas of everyday life where abuse is especially rampant: the media,

gambling, politics, finance, commerce, and of course, advertising."

6. J. Duncan and A. Gross, *Statistics for the 21st* Century (Chicago: Irwin, 1995). Explores "strengths, weaknesses, opportunities, and threats posed by statistical data generated by government bureaus, associations, international agencies, and corporations." It is a useful companion to every text on statistics. Its attitude and spirit are invaluable for all who study, apply and make use of statistical methodology.
7. R. Hooke with J. Liles, *How to Tell the Liars from the Statisticians* (New York: Marcel Dekker, 1983). "Hooke discourses on some 76 topics involving statistics, from drug laws, through smoking and cancer, the cost of living, quality control, political polls and psychological tests, to random digits and computer simulation. He explains basic statistical principles simply and disposes of many common statistical fallacies." (*The American Mathematical Monthly*, Aug.–Sept. 1984)
8. D. Huff with I. Geis, *How to Lie with Statistics* (New York: W.W. Norton, 1954). "This book is a sort of primer in ways to use statistics to deceive. It may seem altogether too much like a manual for swindlers. Perhaps I can justify it in the manner of the retired burglar whose published reminiscences amounted to a graduate course in how to pick a lot and muffle a footfall. The crooks already know these tricks; honest men must learn them in self-defense. "A highly readable, insightful, engaging book on slippery statistics which must be at least in its twelfth printing. Although it's almost half a century since *How to Lie with Statistics* appeared, its illustrations are timeless.
9. A. Jaffe and H. Spirer, *Misused Statistics: Straight Talk for Twisted Numbers* (New York: Marcel Dekker, 1987). "*Misused Statistics* describes sophisticated statistical issues so that nonstatisticians can understand them. . . ." In this it succeeds. It's in the class of books pioneered by Huff's classic, but if I had both books before me and had to choose which to read first, I would go by author in alphabetical order. But then I would continue with this insightful work.
10. R. Reichard, *The Numbers Game: Uses and Abuses of Managerial Statistics* (New York: McGraw-Hill, 1972). "The book focuses not so much on the mechanics or mathematics of a particular technique but rather on what each technique can and cannot do and on its ultimate meaning. Pitfalls and abuses are outlined

in detail and alert the reader to the surprisingly large number of ways in which he can be misled or duped by seemingly accurate statistics."

11. M. Traugott and P. Lavrakas, *The Voter's Guide to Election Polls* (Chatham, N.J.: Chatham House Publishers, 1996). An excellent, highly readable primer on election polls. One of its goals "is to help citizens develop a more critical view of how polls do and don't, yet could and should, contribute to a more informed electorate and a better functioning democracy."
12. M. Wheeler; *Lies, Damn Lies, and Statistics: The Manipulation of Public Opinion in America* (New York: Liveright, 1976; Dell, 1976). "The question of the trustworthiness of the polls goes beyond questions of methodology or careless coverage by the press. So much now depends on the opinion polls—political candidacies, policies, profits—that there is great pressure on the pollsters to manipulate their surveys. It is not hard to rig a poll. . . . This book is about public opinion polls—how they are used and abused."

COMPUTERS TO THE RESCUE?

Computer technology has provided us with a powerful tool for teaching mathematics, at least one dimension of mathematics. Robert Berkman notes [11]:

> Over the past 12 years, I have seen computers make a real difference in how children use and understand information. I have seen children create detailed polished research projects far beyond what I could have produced when I was their age because the labor of editing and rewriting stood in my way.
>
> I have seen children learn more about algebraic equations in two weeks than I did in a year, because the computer freed them from plotting hundreds of points by hand. I have seen children design computer programs portraying natural phenomena that not even my best shoebox diorama could show.

Impressive, and without a doubt important. This was probably the dimension of the computer's application to mathematics which left a lasting impression on Bottom-Line Bob ("But I Used a Computer!" Chapter 10). As Bob put it: "Ramunė, I still don't quite get it. The company spent millions of dollars to update its computer system. I had access to the latest and the best. Why didn't this guarantee success?"

Ramunė's reply provides a perspective for computer usage in general: "The statesman Henry Clay's observation that statistics are no substitute for judgment applies equally well to the computer, Bob. We cannot expect the computer to convert unrealistic assumptions into golden truths. Keep in mind NINO: If nonsense in, nonsense out. I'm sorry to have to have to say that indiscriminate use of computer power has made possible the generation of more nonsense from unrealistic assumptions more quickly than ever before by people having little understanding of what they are doing."

Epilogue

To obtain an overall perspective on the nature of slippery math, its consequences for our quality of life, and what can be done to limit its spread and influence, we return to the seven questions posed in the preface.

1. *What is slippery math and how does it arise?*

Mathematics as a discipline in its own right, separate from its real-world applications, lives up to its reputation for precision. In their own right 1 plus 1 is 2, 8 times 7 is 56, math proof carried out in accord with the standards which define math proof is precise in this respect. One sense of mathematical precision, which we acquire from our education, comes from this dimension.

Our second sense of mathematical precision comes from what we see of math proof, principally from the study of geometry at the secondary school level. Who could remain unimpressed by the sequence of steps in a geometric proof which inexorably lead to the conclusion to be established? Since the setting of geometry has to do with spatial relations whose real world truth may impress us as being "obvious," it is understandable that many of us (Like Jules Warner's Uncle George; see "'Precise' Mathematical Reasoning According to Uncle George," Chapter 8) view math proof as a powerful and precise tool for establishing real world truth.

The slippery dimension of mathematics comes from its applications to real-world problems and situations. Numbers which impress us as being so exact in a mathematical sense often fail to live up to expectations in real world situations. The reasons for this and ways in which it may happen are discussed in Part One. Conclusions obtained by math proof, so seemingly irrefutable, often fall short of expectations of truth in real world settings. The reason for this is the subject of Part Two.

2. *What have been the consequences of slippery math in the public arena?*

Poor decision making, bad legislation, wasted resources, and lost opportunities. For discussion, see Part Three.

3. *How is math employed to sell a point of view?*

The basic idea is to take advantage of suggested respectability by association. If Arthur is seen in the company of the wealthy and powerful, does this not suggest that he, too, is wealthy and powerful? If a point of view is seen in the company of mathematics, which is associated with precision, certainty, and objectivity, does this not suggest that it too has these qualities? The suggestion in both cases is a powerful one, which has often been exploited to gain advantage.

These observations should not be taken to suggest that whenever math is enlisted to support a point of view it is probably an illegitimate use. Far from it; it is a matter of being cautious and distinguishing between the legitimate use of mathematics and its use for show.

The place of math in the spin doctor's bag of tools is further discussed in Chapter 12.

4. *With billions of dollars and endless talk given over to "improving" mathematics education, how could it happen that math could be employed to mislead most of the people most of the time?*

The key phrase is improving mathematics education, which prompts us to ask: Improve mathematics education in what sense? The answer is inherent in the current view of the major objectives of mathematics education: Provide our children with a mathematics education that gives them every advantage in pursuing success in their chosen careers and ensure that the United States maintains supremacy in a modern technologically driven world. (See "Math Education for What?" Chapter 16.) These major objectives have not changed and discussion has focused on how to improve mathematics education to better achieve them. (The new-new math vs. the traditional math debate). This is the dimension that has absorbed billions of dollars.

The mathematics education establishment has not given systematic thought to the broader goal of providing our children with a mathematics education that gives them an understanding of what mathematics as a discipline can do for us and its limitations in applications, what mathematical precision means and its limitations.

One might argue that this goal should be inherent in any mathematics education program. I would agree, but it has not worked out that way in practice. Until this goal is explicitly formulated as a major objective

and seriously addressed we will not see a significant change in public discourse in which numbers and math based arguments could be employed to mislead most of the people most of the time.

5. What reforms in mathematics education are needed to help prepare future generations to effectively participate in a society in which mathematics is increasingly being employed and misemployed in the public arena?

I would suggest that in a general sense the answer follows from the discussion of the preceding question: Identify as a major goal of mathematics education the broad based goal of providing our children with an education which provides them with an understanding of what mathematics can do for us and its limitations, what mathematical precision means and its limitations. If this goal were achieved, students would be positioned to critique the following commonly held math views cited in the introductions to Parts One and Two.

1. A presentation supported by figures is more credible than one that is not.
2. Numbers are neutral. They are not affected by cultural, economic, and political differences that influence people and therein lies their strength. 1 + 1 = 2, for example, is universal for all peoples, countries, economic and political systems.

3. Numbers give weight to a view through the sense of precision they communicate, thereby advancing it to a plane which commands recognition, respect, and acceptance.
4. Mathematical proof is the most reliable means for objectively establishing truth.
5. The precision of mathematical methods (that is, the techniques used to construct a math proof) guarantees unassailable conclusions that serve as pillars of stability and strength in a world besieged by foggy thinking, prejudice, and rampant special interests.
6. Mathematically derived conclusions are indisputable because they are based on deductive logic, which is untainted by bias and ideology.
7. Mathematical methods have the advantage of being certain. No qualified person can resist the truth of a mathematical conclusion properly communicated. The job of communication may be difficult if the solution is complex, but when the communication is competent, agreement is inevitable.

One positioned to critique these views (see Chapters 7 and 13) would be armed with a perspective that would make him or her immune to most forms of slippery math that arise in the conduct of public affairs.

6. *What can we do to make ourselves less vulnerable to being math manipulated by slippery math?*

To address this question, let us first return to the exchange between Bottom-Line Bob and Reflective Ramunė on mathematics ("Which Solution Should Be Implemented?" Chapter 10).

Bob: "What I like most about math is its precision. You don't get 10 sides to a story. You get one answer and that's that; no baloney. Give me the numbers and that ends all arguments; no more beating about the bush."

"Does this mean there's no room in your number world for others who might seek to make sense of numbers," inquired Ramunė.

"What's there to make sense of?" replied Bob. "Numbers speak for themselves."

"But through whose judgment, and how reliable is that judgment?" countered Ramunė.

"My judgment," replied Bob, "that's what Bottom-Line means."

"Yes, but I'm afraid that that may be quicksand masquerading as bedrock," answered Ramunė.

"It turned out to be quicksand in this situation, which is why you're in the predicament you now find yourself."

Bob seeks anchors of certainty in a world awash in relativism and for him, as for many of us, numbers and math arguments provide such anchors. Even after Ramunė explained to him why his view of the nature of mathematical precision was untenable, Bob found it difficult to let go of a mindset that had taken hold of his thinking over a period of many years, a mindset that had been reinforced by his mathematics education.

The most important step we can take to reduce our vulnerability to being math manipulated is to free ourselves from Bob's kind of math mindset. Since this mindset is intimately linked with the seven aforenoted commonly held views of mathematics, it would be advantageous to return to them periodically, critique them ourselves, and return to the critiques offered in the first sections of Chapters 7 and 13. It would then be useful to review the remaining sections of Chapters 7 and 13 along with other sections of this book. Further suggested readings are noted in Chapter 16, Math Education for Public Affairs.

7. What can we do to help policy decision making be reflective of more reliable and well-chosen numbers and math models?

We have to be willing to raise questions and insist on answers from individuals or groups empowered to make decisions. For example:

1. How do you justify the realism of the figures presented in support of the proposal you are advocating?
2. What assumptions underlie the projections you came up with? What evidence do you have in support of their realism?

If no reply or an unsatisfactory reply is given, we must be willing to persist, either as individuals or as part of an interest group. If an atmosphere of good faith prevails, this effort should encourage those invested with decision-making authority to critically examine their figures and assumptions and make appropriate modifications when they are found to be wanting. If an atmosphere of good faith does not prevail, a basis for calling into question proposed decisions will have been established.

References and Readings of Interest

CHAPTER 1

1. G. Becker, "How the Homeless 'Crisis' Was Hyped," *Business Week*, Sept. 12, 1994.
2. T. Weiner, "General Details Altered 'Star Wars' Test," *The New York Times*, Aug. 27, 1993.
3. W. Broad, "Missile Contractor Doctored Tests, Ex-Employee Charges," *The New York Times*, March 7, 2000.
4. W. Broad, "Antimissile Testing Is Rigged to Hide a Flaw, Critics Say," *The New York Times*, June 9, 2000.
5. T. Postol and G. Lewis, "We Can't Tell the Missiles from the Mylar," *The New York Times*, July 7, 2000.
6. E. Sciolino, "Scientist Is Not Subtle in Taking Shots at Missile Shield," *The New York Times*, July 10, 2000.
7. T. Weiner, "Lies and Rigged 'Star Wars' Test Fooled the Kremlin and Congress," *The New York Times*, Aug. 18, 1993.
8. E. Schmitt, "Aspin Disputes Report of 'Star Wars' Rigging," *The New York Times*, Sept. 10, 1993.
9. P. Hilts, "Data on Secondhand Smoke Were Faked, Workers Say," *The New York Times*, Dec. 21, 1994.
10. W. Broad, "Data Tying Cancer to Electric Power Found to Be False," *The New York Times*, July 24, 1999.
11. K. Eichenwald and G. Kolata, "A Doctor's Drug Studies Turn into Fraud," *The New York Times*, May 17, 1999.

CHAPTER 2

1. J. Allen, "The Numbers Are Honest. Honest." *U.S. News & World Report*, March 8, 1999.
2. A. Sloan, "Washington's Math Problem," *Newsweek*, Feb. 1, 1999.
3. D. Stockman, *The Triumph of Politics: How the Reagan Revolution Failed* (New York: Harper & Row, 1986), pp. 123–124.
4. W. Symonds et al., "Tyco: Aggressive or Out of Line?" *Business Week*, Nov. 1, 1999.
5. F. Norris, "At Tyco, Accounting 'Baths' Begin Before the Deals Close," *The New York Times*, Oct. 29, 1999.
6. Cited in Richard Oppel, Jr.'s article, "Buffett Deplores Trend of Manipulated Earnings," *The New York Times*, March 15, 1999. Oppel's source is Berkshire Hathaway's 1998 report.
7. H. Schilit, *Financial Shenanigans* (New York: McGraw-Hill, 1993).
8. I. Preston, *The Tangled Web They Weave: Truth Falsity and Advertisers* (Madison: The University of Wisconsin Press, 1994), p. 99.
9. L. Altman, "Experts See Bias in Drug Data," *The New York Times*, April 29, 1997.
10. G. Kolata, "Safeguards Urged for Researchers," *The New York Times*, April 17, 1997.
11. R. King, Jr., "Medical Journals Rarely Disclose Researchers' Ties," *The Wall Street Journal*, Feb. 2, 1999.
12. K. Eichenwald and G. Kolata, "Drug Trials Hide Conflicts for Doctors," *The New York Times*, May 16, 1999.
13. K. Eichenwald and G. Kolata, "When Physicians Double as Entrepreneurs," *The New York Times*, Nov. 30, 1999.
14. S. Stolberg, "Biomedicine Is Receiving New Scrutiny as Scientists Become Entrepreneurs," *The New York Times*, Feb. 20, 2000.
15. Herodutus, *The History*, translated by D. Grene (Chicago: The University of Chicago Press, 1987), pp. 535–536.
16. Population estimates of the regions of Greece are difficult to come by. A. W. Gomme, *The Population of Athens in the Fifth and Fourth Centuries B.C.* (1993), gives an estimate of 315,000 for Athens and the surrounding region in the age of Pericles (c. 450 B.C.). Estimates for the population of Sparta and the surrounding region vary from 182,000 to 376,000, c. 480 B.C.; see W. Durant, *The Life of Greece* (1939), pp. 74 and 682. E. Semple, *Geography of the Mediterranean Region* (1931), gives an estimate of 110,000 for Corinth, c. 480 B.C.
17. R. J. Newman, "The Bombs That Failed in Kosovo," *U.S. News & World Report*, Sept. 20, 1999.
18. J. Barry and E. Thomas, "The Kosovo Cover-Up," *Newsweek*, May 15, 2000.

CHAPTER 3

1. G. Schneiders and J. Livingston, "Can You Trust the Polls? Well, Sometimes," *The Wall Street Journal*, Feb. 8, 1999.
2. A. Harmon, "Underreporting Found on Male Teen-Ager Sex," *The New York Times*, May 8, 1998.
3. C. Turner et al., "Adolescent Sexual Behavior, Drug Use, and Violence: Increased Reporting with Computer Survey Technology," *Science*, May 8, 1998.
4. G. Sperling, "Me Against the Pollsters," *The Christian Science Monitor*, June 16, 1998.
5. R. Morin, "Public Policy Surveys: Lite and Less Filling," *The Washington Post National Weekly Edition*, Nov. 10, 1997.
6. R. Morin, "Warts and All," *The Washington Post National Weekly Edition*, Oct. 13, 1997.
7. R. Morin, "Keeping a Watch on the Pollsters," *The Washington Post National Weekly Edition*, Nov. 18, 1996.
8. "Hite / ABC Comparison Analysis," news release by ABC for 6:30 p.m., EST, Monday, Oct. 26, 1987.
9. H. Taylor et al., "Touchdown! Online Polling Scores Big in November 2000," *Public Perspective*, March / April 2001, p. 38. Also of interest: A. Greenberg and D. Rivers, "Pioneer Days: The Promise of Online Polling," *Public Perspective*, March / April 2001; R. Morin, "Will Traditional Polls Go the Way of the Dinosaur?," *The Washington Post National Weekly Edition*, May 15, 2000.

CHAPTER 4

1. M. Anderson and S. Fienberg, *Who Counts? The Politics of Census Taking in Contemporary America* (New York: Russell Sage Foundation, 1999).
2. F. Holzman, "How C.I.A. Invented Soviet Military Monster," *The New York Times*, Oct. 3, 1991.
3. F. Holzman, "C.I.A. Error Still Bloats Our Military Budget," *The New York Times*, Feb. 2, 1993. See also Refs. 4 to 6.
4. F. Holzman, "Of Dollars and Rubles," *The New York Times*, Oct. 26, 1979.
5. F. Holzman, "How C.I.A. Concocts Soviet Defense Numbers," *The New York Times*, Oct. 25, 1989.
6. L. Silk, "Economic Scene: Did C.I.A. Distort Soviet Outlays," *The New York Times*, Nov. 17, 1989.
7. J. Brooke, "In Brazil, Slip of the Tongue Makes Campaign Slip," *The New York Times*, Sept. 5, 1994.
8. S. Tefft, "China Is Under Pressure to Clean Up Its Statistics," *The Christian Science Monitor*, June 9, 1994.

9. D. Sanger, "Bad Debt Held by Japan's Banks Now Estimated Near $1 Trillion," *The New York Times*, July 30, 1998.
10. S. Strom, "Japan Assailed for Omitting Data in Growth Calculations," *The New York Times*, May 24, 2000.
11. S. Strom, "Japan Moves to Correct a Deliberate Omission of Data," *The New York Times*, May 25, 2000.
12. C. Clark, *A Critique of Russian Statistics* (London: MacMillan and Co., 1939).
13. H. Shaffer (ed.), *The Soviet Economy: Western and Soviet Views* (New York: Appleton-Century-Crofts, 1963).
14. J. Duncan and A. Gross, *Statistics for the 21st Century* (Chicago: Irwin, 1995).
15. V. Kirichenko, "Return Credibility to Statistics," *Business Economics*, October 1990.
16. M. Gordon, "Moscow Statisticians Accused of Aiding Tax Evasion," *The New York Times*, June 10, 1998.
17. D. Francis, "Budget Fudge," *The Christian Science Monitor*, Aug. 25, 1999.
18. P. Passell, "Economic Scene: Using Creative Accounting to Reduce the Federal Deficit," *The New York Times*, Oct. 26, 1995.
19. F. Hollings, "Beltway Accounting," *The New York Times*, Oct. 9, 1997.
20. A. Sloan, "The Surplus Shell Game," *Newsweek*, Jan. 19, 1998.
21. M. Miller, "Budget Tricks, 1998 Version," *U.S. News & World Report*, Feb. 16, 1998.
22. A. Sloan, "Books Cooked D.C. Style," *Newsweek*, Feb. 16, 1998.
23. A. Sloan, "The No-Count Surplus," *Newsweek*, Nov. 2, 1998.
24. "Rigging the Numbers," *The New York Times*, June 15, 1998.
25. L. V. Kantorovich, *Mathematical Methods of Organizing and Planning Production*, Leningrad University, 1939. For an English translation, see *Management Science*, vol. 6, no. 4 (July 1960), pp. 363–422; or V. S. Nemchinov, ed., *The Use of Mathematics in Economics* (Cambridge, Mass: MIT Press, 1964).
26. D. Card and A. Krueger's studies are included in their book *Myth and Measurement: The New Economics of the Minimum Wage* (Princeton, NJ: Princeton University Press, 1995).
27. A. Blinder, "The $5.15 Question," *The New York Times*, May 23, 1996. A sample of the attacks launched against the Card-Krueger study is provided by Richard Berman's reply to Blinder's letter [28]. Berman is executive director of the Employment Policies Institute, an interest group for restaurant companies.
28. R. Berman, "Wage Study Was Based on Faculty Research" (letter), *The New York Times*, May 28, 1996.

CHAPTER 5

1. A. Bryant, "A Different Gauge for Rating Airlines," *The New York Times*, March 7, 1995.

2. C. Whalen, "Is the Fed Fighting a Phantom Menace?," *Business Week*, May 29, 2000.

3. S. Cecchetti, R. Chu, and C. Steindel, "The Unreliability of Inflation Indicators," *Federal Reserve Bank of New York: Current Issues in Economics and Finance*, April 2000. Also of interest: L. Uchitelle, "Epitaph for a Rule that Just Won't Die," *The New York Times*, July 30, 2000; D. Francis, "Why Money Supply Deserves More Fed Attention," *The Christian Science Monitor*, July 31, 2000.

4. "Up Front: Deep Thinkers," *Business Week*, Dec. 20, 1999.

5. J. Davidson and W. Rees-Mogg, *The Great Reckoning*, rev. ed. (New York: Touchstone Books, 1994).

6. R. Eisner, *The Misunderstood Economy* (Boston: Harvard Business School Press, 1994).

7. R. Kuttner, *The End of Laissez Faire: American Economic Policy After the Cold War* (New York: Random House, 1991).

8. J. Judis, "Embarrassment of Riches," *The American Prospect*, June 19-July 3, 2000.

9. R. Kuttner, "What's Wrong with Paying Off the National Debt," *Business Week*, May 15, 2000.

10. F. Thayer, "Reducing Debt Has a Depressing History" (letter), *The New York Times*, May 26, 1995. Also of interest in connection with the history of the national debt: J. Gordon, *Hamilton's Blessing: The Extraordinary Life and Times of Our National Debt* (Walker & Co., 1997).

CHAPTER 6

1. R. Morin, "Getting Personal with Political Parties," *The Washington Post National Weekly Edition*, Sept. 26, 1994.

2. G. Casper, letter to J. Fallows, ed., *U.S. News & World Report*, Sept. 23, 1996. With the permission of Casper and Fallows, this letter has entered the public domain.

3. J. McKinley, "Student Group Attacks U.S. News College Guide," *The New York Times*, Nov. 6, 1996.

4. W. Honan, "Stanford Posts Internet Alternative to U.S. News College Guide," *The New York Times*, April 27, 1997.

5. P. J. LeBlanc, "College Ranking Time: It's Like Finals," *The Christian Science Monitor*, Oct. 21, 1997.

6. L. Johannes and J. Welsh, "Law Schools Assail Magazine's Ratings," *The Wall Street Journal*, Feb. 19, 1998.

7. N. Lehmann, "Rewarding the Best, Forgetting the Rest," *The New York Times*, April 26, 1998.

8. A. Kuczynski, "'Best' List For Colleges By U.S. News Is Under Fire," *The New York Times*, Aug. 20, 2001.
9. A. Graham and N. Thompson, "Broken Ranks," *The Washington Monthly*, Sept., 2001.
10. J. Fellows, "The Great College Hustle," *Atlantic Monthly*, Aug., 2001.
11. B. Duffy, "'Best' Colleges List" (letter), *The New York Times*, Aug. 23, 2001.
12. "A Reality Check for College Rankings" (editorial), *The New York Times*, Aug. 27, 2001.
13. D. Streitfeld, "Whose Top 100 Is It, Anyhow?" *The Washington Post National Weekly Edition*, Aug. 10, 1998. Also of interest: A. de Botton, "Great Books, Read and Unread," *The New York Times*, July 22, 1998; F. Rich, "Who Chose 'The Magus'?" *The New York Times*, Aug. 8, 1998.

CHAPTER 7

1. R. Baker, "The Observer: Uncle Pete? Perot?," *The New York Times*, July 11, 1992.
2. A. Mitchell and F. Bruni, "Spotlight Turns on Ugly Side of Politicking," *The New York Times*, Feb. 11, 2000.
3. C-Y Cheng, "Letter to the Editor," *The New York Times*, June 3, 1999.
4. J. Leo, "No, Don't Give Me a Number!" *U.S. News & World Report*, May 14, 1990.
5. G. Will, "Lies, Damned Lies and . . . ," *Newsweek*, March 29, 1999.
6. C. Jones, "Test Scores Show Gaps by Ethnicity," *The New York Times*, July 8, 1994.
7. A. B. Jeffries, "Schools Can't Give Students What They Don't Get at Home," *The New York Times*, July 20, 1994.
8. D. Rosenbaum, "Under the Hood with Perot's Data," *The New York Times*, August 21, 1996.
9. D. Falkner, *The Last Yankee: The Turbulent Life of Billy Martin* (New York: Simon & Schuster, 1992).
10. G. Will, "Paranoid in Pinstripes," *The New York Times Book Review*, April 5, 1992.
11. J. Jagger, "Why Patriot Didn't Work as Advertised," *The New York Times*, June 9, 1991.
12. E. Marshall, "Patriot's Scuds Busting Record Is Challenged," *Science*, May 3, 1991.
13. M. Anderson, "The Reagan Boom—Greatest Ever," *The New York Times*, Jan. 17, 1990.
14. W. Leontief, "We Can't Take More of this Reagan Boom," *The New York Times*, Feb. 4, 1990.

15. "A Poll of AIDS Doctors Finds That Many Help in Suicides," *The New York Times*, Feb. 3, 1997.

CHAPTER 8

1. H. Herring, "Where There's Smoke, There's Outlay," *The New York Times*, April 27, 1997.

CHAPTER 11

1. D. Stockman, *The Triumph of Politics: How the Reagan Revolution Failed* (New York: Harper & Row, 1986), pp. 92, 98.
2. L. Barrett, *Gambling with History: Ronald Reagan in the White House* (New York: Doubleday, 1983), p. 140.
3. For details on this model, see R. D. Buzzell, *Mathematical Models and Marketing Management* (Cambridge, Mass: Harvard University Press, 1964), chap. 5.
4. P. Kotler, "Computerized Media Planning: Techniques, Needs and Prospects," *Occasional Papers in Advertising* (Urbana, IL: American Academy of Advertising, 1965).
5. F. Bass and R. Lonsdale, "An Exploration of Linear Programming in Media Selection," *Journal of Advertising Research*, vol. 3, no. 2 (May 1966), pp. 179–188. For further discussion of linear programming see, for example: W. J. Adams, *Fundamentals of Mathematics For Business, Life and Social Sciences* (Englewood Cliffs, NJ: Prentice Hall, Inc., 1979), Ch. 6.
6. G. Epstein, "Is it Really Reasonable to Assume that the Newest Nobelist Deserved the Prize?" *Barron's*, Oct. 19, 1998.
7. *The Economist*, "The Future of Economics," March 4, 2000. Also of interest: M. Weinstein, "Students Seek Some Reality Amid the Math of Economics," *The New York Times*, Sept. 18, 1999. J. Galbraith, "How the Economists Got It Wrong," *The American Prospect*, Feb. 14, 2000.
8. "The New Economy: Work in Progress," *The Economist*, July 24, 1999.
9. "Readjusting the Lens," *The Economist*, Nov. 20, 1999.
10. M. Mandel, "How Fast Can this Hot-Rod Go," *Business Week*, Nov. 29, 1999.
11. L. Uchitelle, "A Clearer View of the Economy," *The New York Times*, Oct. 29, 1999.
12. M. Mandel, "The Spoils of the New Economy Belong to High Tech," *Business Week*, Aug. 16, 1999.
13. M. Mandel, "Financial Services: The Silent Engine," *Business Week*, Dec. 21, 1998.

14. L. Uchitelle, "Measuring Productivity in the 90's: Optimists vs. Skeptics," *The New York Times*, Aug. 2, 1997.
15. R. E. Peierls, *The Laws of Nature* (New York: Charles Scribner's Sons, 1956), p. 17.

CHAPTER 12

1. "Why 300 Economists, 136 Newspaper Editors and 6 Presidents Support NAFTA. NAFTA Works for Everyone," *The New York Times*, Oct. 24, 1993.
2. G. Hufbauer and J. Schott, *NAFTA: An Assessment* (Washington, D.C. : Institute for International Economies, 1993). The Hufbauer-Schott model predicts the creation of 170,000 jobs in NAFTA's first year. The 200,000 figure for jobs created was obtained by rounding up the 170,000 figure, according to David Walter, the U.S. Trade Representative's chief economist. See T. Lee, *False Prophets, The Selling of NAFTA* (Washington, DC: Economic Policy Institute, 1995; p. 11); also see P. Cooper and L. Wallach, *NAFTA's Broken Promises: Job Creation Under NAFTA* (Washington, DC: Public Citizen, September 1995, p. 7).
3. "The NAFTA Effect: When Neighbors Embrace," *The Economist*, July 5, 1997.
4. R. Pear, "G.O.P. Suggests Smaller Benefit Adjustments," *The New York Times*, May 11, 1995.
5. "Toward a More Accurate Measure of the Cost of Living," Final Report to the Senate Finance Committee from the Advisory Commission to Study the Consumer Price Index; Dec. 4, 1996.
6. J. Madrick, "The Cost of Living: A New Myth," *The New York Review of Books*, March 6, 1997. Madrick presents an informative review of and commentary on the Boskin Commission's report and three other books concerning the CPI.
7. D. Baker, "The Inflated Case Against the CPI," *The American Prospect*, winter 1996.
8. D. Francis, "Fixing the Inflation Index—But Is It Really Broken?," *The Christian Science Monitor*, March 6, 1997.
9. D. Francis, "Poking Holes in the C.P.I. Balloon," *The Christian Science Monitor*, March 14, 1997.
10. P. Passell, "Some Experts Say Inflation Is Understated," *The New York Times*, Nov. 6, 1997.
11. J. Berry, "A Numbers Game Played for High Stakes," *The Washington Post National Weekly Edition*, March 17, 1997.
12. D. Baker and M. Weisbrot, *Social Security: The Phony Crisis* (Chicago: The University of Chicago Press, 1999, p. 3). I did not see details about the poll—how the poll was carried out, sample of questions posed, margin of error, etc.—and per chap. 3, nos. 2, 3, this makes me uncomfortable. Peter Hart

has a solid reputation and it's reasonable, I believe, to view these results as in reality's ballpark.

13. A. Bernstein, "Social Security: Is the Sky Really Falling?, *Business Week*, Feb. 10, 1997.
14. A. Bernstein, "Social Security: Go Refigure," *Business Week*, Feb. 8, 1999.
15. "What Social Security Crisis" (editorial), *Business Week*, Nov. 30, 1998. This editorial was reproduced in *The New York Times*, under Business Week Commentary, Nov. 21, 1998 (A. 15).
16. H. Aaron, "The Myths of the Social Security 'Crisis,'" *The Washington Post National Weekly Edition*, July 29, 1996.
17. D. Baker and M. Weisbrot, cited in Ref. 12.
18. P. Coy, "Social Security: Let It Be," *Business Week*, Nov. 30, 1998.
19. R. Eisner, *The Great Deficit Scares: The Federal Budget, Trade, and Social Security* (New York: The Century Foundation Press, 1997).
20. R. Eisner, *Social Security: More, Not Less* (New York: The Century Foundation Press, 1998).
21. G. Koretz, "Stay Loose on Social Security," *Business Week*, March 29, 1999.
22. "Is There a Social Security Crisis? Sam Beard Debates Theodore R. Marmor & Jerry L. Mashaw," *The American Prospect*, January–February 1997.
23. J. Madrick, "Social Security and Its Discontents," *The New York Review of Books*, Dec. 19, 1996.
24. A. Munnell and R. K. Weaver, "Social Security's False Alarm," *The Christian Science Monitor*, July 19, 2001.
25. J. D. McKinnon, "Bush Commission Begins to Make Case That Social Security Must Be Overhauled," *The Wall Street Journal*, July 20, 2001.
26. R. W. Stevenson, "Panel Argues for Changing Social Security," *The New York Times*, July 20, 2001.
27. P. Krugman, "2016 and All That," *The New York Times*, July 22, 2001.
28. A. Goldstein, "Bitter Words over Social Security," *The Washington Post National Weekly Edition*, July 30–Aug. 12, 2001.
29. P. Krugman, "Nothing for Something," *The New York Times*, Aug. 8, 2001.
30. P. Krugman, "Fabricating a Crisis," *The New York Times*, Aug. 21, 2001.

CHAPTER 13

1. D. Barry, "City Study Sees Big Revenues if Yankees Play in Manhattan," *The New York Times*, April 22, 1998.
2. P. Passell, "Local Payoff on a Stadium Is Uncertain," *The New York Times*, April 30, 1998.
3. "Focusing on the Surplus," *The New York Times*, Feb. 8, 2000.

4. R. Pear, "Budget Office Says Clinton Underestimated Cost of Drug Plan," *The New York Times*, July 23, 1999.
5. "Why 300 Economists, 136 Newspaper Editors and 6 Presidents Support NAFTA: NAFTA Works for Everyone," *The New York Times*, Oct. 24, 1993.
6. R. Kuttner, "The Fix Is in on the CPI," *The Washington Post National Weekly Edition*, March 17, 1997.
7. J. Madrick, "The Cost of Living: A New Myth," *The New York Review of Books*, March 6, 1997.

CHAPTER 14

1. W. Styron, *Darkness Visible: A Memoir of Madness* (New York: Vintage, 1990). Also see "Prozac Days, Halcion Nights," *The Nation*, Jan. 4/11, 1993.
2. G. Kolata, "Maker of Sleeping Pill Hid Data on Side Effects, Researchers Say," *The New York Times*, Jan. 20, 1992.
3. F. Norris, "Market Place," *The New York Times*, Aug. 1, 1997.
4. M. Petersen, "Board Gives Final Approval to New Rule on Derivatives," *The New York Times*, June 2, 1998.
5. A. Benedi, "A Response to the Makers of Tylenol; and an Important Safety Message to Consumers," *The New York Times*, March 20, 1996.
6. D. Grady, "Tylenol Overdose Linked to Acute Failure of Liver," *The New York Times*, Oct. 16, 1997.
7. D. Grady, "Study Says Thousands Die from Reaction to Medicine," *The New York Times*, April 15, 1998.
8. C. Kalb, "When Drugs Do Harm," *Newsweek*, April 27, 1998.
9. S. Fried, *Bitter Pills: Inside the Hazardous World of Legal Drugs* (New York: Bantam Books, 1998).
10. N. Shute, "Pills Don't Come with a Seal of Approval," *U.S News & World Report*, Sept. 29, 1997.
11. G. Kolata, "The F.D.A. Approves a Drug. Then What?" *The New York Times*, Oct. 7, 1997.

CHAPTER 15

1. S. Faludi, "Statistically Challenged," *The Nation*, April 15, 1996.
2. M. Gardner, "The Power of Statistics to Affect Lives—Even When They're Wrong," *The Christian Science Monitor*, May 2, 1996.
3. K. Eichenwald and G. Kolata, "Drug Trials Hide Conflicts for Doctors," *The New York Times*, May 16, 1999.
4. B. Carter, "Watching the Watchers," *The New York Times*, March 10, 1997.

5. B. Carter, "Networks Battle Nielson as Young Viewers Turn Up Missing," *The New York Times*, Dec. 21, 1998.
6. F. C. Thayer, "Do Balanced Budgets Cause Depressions?" *The Washington Spectator*, Jan. 1, 1996.
7. "The Unmeasurable Lightness of Being," *The Economist*, Nov. 23, 1996.
8. "Damned Lies: Economic Statistics Are in a Bad Way" (editorial), *The Economist*, Nov. 23, 1996.
9. E. Ehrlich, "The Downside of Bad Data," *Challenge*, March–April 1997.
10. M. Mandel, "The Real Truth About the Economy," *Business Week*, Nov. 7, 1994.
11. M. Weidenbaum, "Fund the Number Crunchers," *The Christian Science Monitor*, Sept. 16, 1999. Also of interest: H. Gleckman, "On Congress' Hit List: Crucial Business Data," *Business Week*, Sept. 13, 1999; S. Strong, "The Link Between Quality of Data and Quality of Life," *The Christian Science Monitor*, Jan. 30, 1997.

CHAPTER 16

1. L. Walters, "Rebel Parents See "'New-New Math' as a Big Minus," *The Christian Science Monitor*, July 2, 1997.
2. L. Cheney and T. Romberg, "Dialogue: Creative Math, or Just 'Fuzzy Math'?, Mediocre Is Not Good Enough," *The New York Times*, Aug. 11, 1997.
3. Letters in reply to the Cheney-Romberg Dialogue, *The New York Times*, Aug. 17, 1997.
4. J. Steinberg, "California Goes to War Over Math Instruction," *The New York Times*, Nov. 27, 1997.
5. Letters in reply to "California Goes to War Over Math Instruction," *The New York Times*, Dec. 1, 1997.
6. D. Hunsaker, "Ditch the Calculators," *Newsweek*, Nov. 3, 1997.
7. J. Steinberg, "Clashing Over Education's One True Faith" *The New York Times*, Dec. 14, 1997.
8. "Scholars Band Together to Oppose Mathland," *The Christian Science Monitor*, Nov. 30, 1999.
9. A. MacLeod, "Trash the Calculator, It's Back to Basics in Britain," *The Christian Science Monitor*, Jan. 26, 1998.
10. A. Hartocollis, "Math Teachers Back Return to Education in Basic Skills," *The New York Times*, April 13, 2000.
11. R. Berkman, Letter, *The New York Times*, May 21, 1996.

Index